E:CODS

Emergence: Complexity & Organization
**An International Transdisciplinary Journal of
Complex Social Systems**
VOLUME 14, Number 4, 2012
Complexity & Public Policy

Published and distributed by:

EMERGENT™
PUBLICATIONS
3810 N 188th Ave
Litchfield Park, AZ 85340, USA

i

E:CO:E

Emergence: Complexity & Organization
An International Transdisciplinary Journal of Complex Social Systems

Editors-in-Chief
PETER ALLEN, *Complex Systems Research Centre, Cranfield University, Bedford, UK*
JEFFREY GOLDSTEIN, *Adelphi University, Garden City, NY, US*
DAVID SNOWDEN, *Cognitive Edge, UK*

Founding Editor of Emergence
MICHAEL LISSACK, *ISCE, Naples, FL, USA*

Managing Editor and Production Editor
KURT RICHARDSON, *Emergent Publications, Litchfield Park, AZ, US*

Graphic Design
MARSHALL CLEMENS, *Idiagram, Boston, MA, USA*

Subject Editors
Innovation & Networks: PIERPAOLO ANDRIANI, *Durham University, UK*
Organizational Knowledge & Learning: ELENA ANTONACOPOULOU, *University of Liverpool, UK*
Strategy, Leadership & Change: DOUGLAS GRIFFIN, *University of Hertfordshire, UK*
Economics & Markets: STAN METCALFE, *University of Manchester, UK*
Philosophy: RIKA PREISER & JANNIE HOFMEYR, *Department of Philosophy, Stellenbosch University, ZAF*
Methodology: PEDRO SOTOLONGO, *Instituto de Filosofia de La Habana, CUB*

Subscription prices (2012, volume 14, 4 issues) not including postage and handling: US$1295.00 Corporate/US$795.00 Institutional/US$99.95 Individual (complementary print on request and electronic access to all previous issues). Members of the ISSS can subscribe to this journal at a concessionary rate. They must declare that the subscription is for their own private use, it will not replace any institutional subscription, and it will not be put at the disposal of any library. Subscriptions should be sent directly to the Emergent Publications, 3810 N 188th Ave, Litchfield Pk, AZ 85340, USA, or to any subscription agent.

VOLUME 14, Number 4, 2012
Special Issue
Complexity and Public Policy

Contents

Philosophy Section

Classic Paper Section

Forum Section

Editorial

Editorial: Innovative Public Policy—The Role of Complexity Science

Simone Landini & Sylvie Occelli
IRES Piemonte, Istituto di Ricerche Economico Sociali del Piemonte, ITA

In the past decade, evidence has been mounting that conventional approaches to policy making have to change. Concerns about sustainability, climate changes and the cascade of effects engendered by the recent economic crisis, just make the evidence even more striking.

Indeed, the acknowledgement that a change of paradigm is needed to cope with the complexity of human organizations and behaviors has been around for some time by now[1].

Since the seventies, the tenets of conventional socioeconomic approaches, such as agents' perfect rationality, market equilibrium and linearity in the cause-effect relationships and between micro and macro behaviors, have been scrutinized. The advancements in system thinking and their cross-fertilization among several study domains, such as biology, computer science, physics, mathematics and cognitive science, contributed to the formulation of alternative tenets offering insights into the complexity features of human activity systems[2].

Science provides an argument that human complex systems a) are made of interacting heterogeneous individual decision-makers which self-organize; b) form building blocks, which coevolve with others, as well with the hosting environment; c) reveal unexpected and emergent features, and; d) the functioning of human activity systems is confronted with certain collective issues which are intrinsically wicked, and therefore, un-tameable.

Two distinct views, notwithstanding the multi-disciplinary efforts undertaken, have been retained in developing complexity oriented policy approaches: a) a

---

1. The science of complexity has also attracted the attention of the European Commission which since FP6 has launched a number of initiatives aimed at investigating its applications in a wide range of scientific domains, see Weisbuch & Salomon (2007) and, more recently OECD(2009).

2. Scientists in physics, geography, computer science, sociology, economy and management, have engaged to provide insights into how human behaviours, organizations, social norms and cultural institutions encroach in and superimpose one another, generating unexpected outcome and unanticipated effects. An overview from different perspective is offered by Lane *et al.* (2009); Helbing (2009); Nicolis & Nicolis (2007); Dennard *et al.* (2008); Mitleton-Kelly (2003); Chu *et al.* (2003); Morin & Le Moigne (1999).

so called external view, mainly endorsed for scientific quest, aimed at understanding the dynamics of the study system and enhancing its steering capability (a long standing issue in the earlier cybernetics studies); b) an internal view, generally adopted by policy stakeholders, concerned with the practices of policy making and namely with the compliances imposed by institutional commitments (strategy design and sector programmes) and procedural tasks (problem definition, implementation of alternatives, monitoring, and evaluation).

Both views had an important role in delving into the range of complexity facets underpinning policy endeavours. Their implications however have been contrasting.

According to the former, addressing the processes of complex socio-economic situations is an essential pre-requisite to devise more effective policy-making. As also OECD (2009) puts it "beyond concepts, tools and methods, complex systems science offers some new ways to think about policy making. It focuses attention on dynamic connections and evolution, not just on designing and building fixed institutions, laws, regulations and other traditional policy instruments" (p.13).

According to the latter, the exposed wickedness of certain situations is usually focussed at interpreting the failures of policy programs or the inability to gauge social priorities which, in both cases, tend to create a sense of un-capacitated actions. In most situations, when the complexity dimensions are grasped their understanding is often metaphorically laden and perceived to have no applications in real practices[3].

The need to bridge the two views is a fundamental requirement to align policy with a transformation occurring in society and deliver innovative public policies (Inguaggiato & Occelli, 2012).

From a different vantage point, an exhortation comes also from OECD (2011), which maintains that to govern means to share *authority, ideas and information* with rivals as well as partners. It entails to encompass not only public organizations and institutions (government), but also methods and instruments for governing (governance).

The investigation and reshaping of *the system of relationships* which, through the two views, link government and governance are the main challenges for leveraging complexity thinking in policy activity[4].

---

3. Occasionally this understanding may urge emergency actions, or result in the introduction of additional procedural controls to manage the complexity situations. When persisting over time, this sense of un-capacitated actions is likely to frustrate any attempts for improvements even causing a disengagement in policy making.

4. The challenges have been recently taken up in the policy oriented satellite meetings, organized at the European Conference of Complex Systems in 2010, 2011 and 2012 (www.complexssociety.eu). The GSDP, FP7 funded coordination and support action is currently de-

The endeavour is not only a matter of informed-based design, but also a context dependent pathway which has to be undertaken on a multiple scale in a forward looking perspective, by: a) engaging the system's actors, namely government organizations, societal partners, and scientists; b) leveraging the existing information; c) favouring the information sharing, knowledge creation and learning, and; d) creating institutional and administrative conditions for actions (Occelli & Semboloni, 2011).

These topics form a common scope of interest in the contributions presented in this *E:CO* issue. Of course, depending on the specific case study and authors' scope, a certain variety of pathways are suggested, which can support the endeavour.

One possible way to grasp this variety is to consider the contributions as belonging to a continuum, ranging from an interest to provide insights into forms and patterns of these relationships, to a more focussed attention towards their overall reshaping.

Keeping in mind that each contribution is a mix of the two, in the following we give emphasis to certain aspects which, taken together, reflect the richness of the pathways suggested by the presenters.

The acknowledgement that conventional policy practices (including planning and management) based on an understanding of social systems as monolithic silo-type entities are ineffective is the main focus of attention in both papers by Morçöl and Verweij.

Morçöl's discussion about urban sprawl is a case in point. He shows that the containment measures adopted in the studied areas to deal with their spatial development only partially succeeded in dealing with sprawl phenomena. Arguments are put forward that the observed failures can be interpreted as an inability to acknowledge the complexity of spatial growth. A suggestion is made that complexity based analytic tools, such as multi-agent simulation models, can help urban agents avoid top-down imposed plans and undertake activities allowing themselves to design their own plans. He reminds us however, that self-organization, per se, is not a magic recipe which can avoid wicked policy issues. Whenever collective choices are involved, in fact, conflicts are likely to emerge as far as goals and resource distribution are concerned.

This recalls the existence of the multiple interactions shaping a policy situations which needs to be made explicit. This is the main focus in Verweij's paper which analyses the interactions of management with a complex institutional context, in the realization of an important transport project in the Netherlands. The managers' roles, in balancing two apparently contrasting dimensions of fragmentation and integration in the project development are investigated, on the basis of

---

veloping a research program for the study of global systems in an ongoing dialogue with decision makers (www.globalsystemdynamics.eu).

a grounded analysis based on interviews and secondary documents. The study shows that while both dimensions are necessary in developing projects; their modulation depends on the context. Flexibility and adaptive capacity are recognized as main requirements in this respect.

The role that complexity thinking can offer in supporting policies is specifically addressed in Daniels's paper. It builds upon three case studies dealing with US foreign policy regarding the Middle East and distills in the form of propositions some general features a system based policy paradigm can take advantage of:

1. The acknowledgement of culture as an emergent feature of a policy structure;
2. The bridging of gaps caused by compartmentalized policy structures;
3. The engagement in a double-loop learning process: openness to learning from errors and willingness to reflect on the underlying assumptions;
4. The minimization of communication and authority disconnects associated with the power distance among groups, and;
5. The adoption of a conceptualization of social systems as evolutionary.

The implications of these propositions do not only include intangible benefits associated with a different way of looking at the world, but also ways to access their advantages. Some examples in this regard are mentioned for the US foreign policy, and the topic will certainly deserve further attention in other policy domain.

Lehmann's study of violent crime in Rio de Janeiro is an exemplary case of how an understanding of the phenomenon as a complex adaptive system, enabled institutions to counteract the problem. Notably, by this understanding, a shift was made from top-down policy actions based on confrontation or containment, to initiatives resulting from locally based social activity, involving new police departments (the permanent Pacifying Police Units) and the communities. By establishing new forms of relationships between police and population, the working of the Pacifying Police Units made considerable progress in making some of Rio's most violent shanty towns more secure. A surprising finding from the study is that such understanding was not based on any theoretically informed complexity based framework. Instead it resulted as a logical response to the failures of the previous safety policies. This clearly shows that there is a potential to adopt the complexity principles in public policy. However, an effort should be made to increase awareness that to adhere to these principles should not be seen as a threat, but as empowerment for the working of the actors engaged.

MacGillivray and Gallagher's paper presents a successful complex policy initiative in Canada: the Lake Simcoe Plan. This policy initiative involved a "comprehensive understanding of human systems" as the approach enabling diverse de-

E:CO Vol. 14 No. 4 2012 pp. vii-xiii

cision-making actors to provide effective interventions. By involving complexity theory perspective and conceptual tools, the authors try to understand if participants to a policy-making process perceive the complexity in which they were embedded. Hence, this paper puts forward a twofold vision from the inside and outside. If the policy-making activity pertains to institutions, the policy-making process involves society; this paper explicitly points out how to bring society into policy-making. The two sub-systems are circumscribed by boundaries which are not elements of separation. On the contrary, they are structural elements putting sub-systems in contact and interaction, in a coevolution. Rather than being barriers, boundaries are communication membranes, allowing for the information exchange which, within the frame of an adaptive management strategy, leads to the comprehension of needs and means and induces the success of the planning initiative. Paradoxically, citizens focused on ordered scientific data in order to gain entry to policy processes and thereby increase social complexity.

An example of a fully developed complexity approach to healthcare system is provided by Sturmberg. It addresses importance of understanding the interrelationships between policy and practice in order to overcome the fragmentations observed in the health domain. Central to the approach is the emphasis on the 'personal experience of health' as the 'core values' of a seamlessly integrated health system. These are at the basis of the structural requirements of healthcare system as a complex adaptive system. The 'vortex' metaphor is used to describe how the system can reorganize by adhering to its core values. This perspective proposes connections between the micro-network (patients) with the macro-network (policy-makers) through a meso-level network (health organizations and communities). The relationships among the system's levels make it possible to learn about health care needs and resource.

Two papers develop theoretical reasoning by using a philosophical language. They both highlight the needs for new policy making paradigms and put forward the idea of a system capable of looking at itself either from inside or outside.

From the point of view of ontological perspective applied to complex human systems, Poli's contribution develops the notion of anticipatory governance. This paper develops the idea that although human beings are well equipped with anticipatory capabilities, the systems they constitute are not equipped in the same way. Anticipatory governance then, turns out to be an ensemble of relationships linking practices, hierarchical structures and differentiated (cultural) perspectives of the world. A stimulating finding is that to manage complexity, uncertainty, disequilibrium and risk, in order to improve societies' behavior and development, human systems cannot be limited to learn from the past. Rather, they should also be able to shape their own futures by thinking and re-thinking themselves while gathering signals on new needs and attitudes.

Fu and Bergeon draw upon the Tao's philosophy in dialogue with Nature and modern physics phenomenology to introduce a new paradigm to view systems as ensembles of activating forces, continuously transmuting energy-being from the past to the present. At-the-moment emergence is what can (must) be known to envisage future effects of policy making. The "actuality of the past" emerges into the present being that informs "potentiality of the future" in a continuous flow. According to the "synchronizing with Nature perspective" from a Tao philosophical perspective, a system is found to be complex, and one of the complexity traits is recognized is the gap between how the system perceives itself, for instance how policy making institutions look at the society, and how its constituents perceive the system they are living in. That is what the society needs, asks for, and expects from policy making activity. The Tao Complexity Tool proposes shifting the policy making activity paradigm from a discrete sequence of problem solving steps into a continuous flow of shared information to discern problems and then solve them.

Complex system sciences provide a well-structured and formalized environment to manage complexity in human systems. Although promising, current applications tend to overlook their implications in context dependent policy practices.

Paradoxically, societies are facing the growth of a new type of needs induced by technical and scientific progress. While scientific development is progressing at a speedy pace, societies and institutions are lagging behind (van del Leeuw, 2012). There is a risk that delays between science and society will increase gaps between theory and practice, means and ends, needs and resources.

This calls for new research fields to address new and old problems.

It seems unreasonable to slow down what is fast, especially when this has a lot of potential for improving society's wellbeing. The only reasonable alternative is to boost what is behindhand, having a look at the future. This is expected to be one major commitment for policy-makers to engage themselves in such undertaking.

The papers of this special issue share a common message: what is missing is not the development of complexity theory for human activity systems but a better understanding of complexity acting in society.

References

Chu, D., Strand, R., and Fjelland, R. (2003). "Theories of complexity: Common denominators of complex systems," *Complexity*, ISSN 1099-0526, 8(3):19-30.

Dennard, L.F., Richardson, K.A., and Morçöl, G. (eds.) (2008). *Complexity and Policy Analysis: Tools and Concepts for Designing Robust Policies in a Complex World*, ISBN 9780981703220.

Helbing, D. (2009). "Managing complexity in socio-economic systems," *European Review*, ISSN 1062-7987, 17(2): 423-438.

Inguaggiato, C. and Occelli, S. (2012). "Policy making in an information wired environment: re-aligning government and governance relationships by complexity thinking," paper presented at the Satellite Meeting on *Complexity in the Real World*, ECCS 2012, Brussels, September 5-6.

Lane, D., van der Leeuw, S, Pumain, D., and West, G. (eds.) (2009). *Complexity Perspectives in Innovation and Social Change*, ISBN 9781402096624.

Mitleton-Kelly, E. (2003). "Ten principles of complexity and enabling infrastructures," in E. Mitleton-Kelly (ed.), *Complex Systems & Evolutionary Perspectives of Organizations: The Application of Complexity Theory to Organizations*, ISBN 9780080439570, pp. 23-51.

Morin, E. and Le Moigne, J.L. (1999). *L'Intelligence de la Complexité*, ISBN 9782738480859.

Nicolis, G. and Nicolis, C. (2007). *Foundations of Complex Systems: Nonlinear Dynamics, Statistical Physics, Information and Prediction*, ISBN 9789812700438.

Occelli, S. and Semboloni, F. (2011). "Bridging expert and lay knowledge in policy making activities: Which role(s) for models?" paper presented at the Satellite Meeting on *Policy Modeling*, ECCS 2011, Vienna, September 14-15.

OECD, (2009). "Applications of complexity science for public policy: New tools for finding unanticipated consequences and unrealized opportunities," http://www.oecd.org/science/scienceandtechnologypolicy/43891980.pdf.

OECD, (2011). Government at a Glance 2011, ISBN 9789264096578.

van der Leeuw, S. (2012). "Global systems dynamics and policy: Lessons from the distant past," Complexity Economics, ISSN 2210-4275, 1(1): 33-61.

Weisbuch, G. and Salomon, S. (2007). *Tackling Complexity in Science: General Integration of the Application of Complexity in Science*, ISBN 9789279055041.

Urban Sprawl And Public Policy: A Complexity Theory Perspective
E:CO Issue Vol. 14 No. 4 2012 pp. 1-16

Urban Sprawl And Public Policy: A Complexity Theory Perspective

Göktuğ Morçöl
School of Public Affairs, Pennsylvania State University at Harrisburg, USA

The epistemological and methodological implications of complexity theory for understanding urban sprawl are discussed. It is argued that urban spatial forms, such as sprawl, emerge from nonlinear, self-organizational, and dynamic urban processes. Because of this, there cannot be a universal theory of sprawl and each case should be investigated within its context. The micro–macro problem provides the conceptual grounding for these investigations. Agent-based simulations can be used to investigate the micro–macro transformations in urban systems. Implications of complexity theory for understanding the role of urban policies are discussed.

Introduction

Urban sprawl has concerned urban scholars, planners, and policymakers in the United States for decades now. An increasing number of scholars, planners, and policymakers in Europe, Australia, and Canada have turned their attentions to this issue in the last couple of decades. The definitions, causes, and consequences of sprawl have been debated in the urban planning and policy literatures since the 1930s in the United States (Bruegmann, 2001). This literature indicates that sprawl is a complex theoretical policy problem, with multiple and even contending definitions and conceptualizations. These contending conceptualizations are behind the heated policy debates on urban sprawl. In this paper I will review the conceptualizations in the literature, illustrate the complexity of the sprawl phenomenon with examples, and propose a complexity theory-based alternative conceptualization.

There are roughly two sides in the theoretical and policy debates on sprawl. There are those policy and planning theorists and researchers who hold sprawl responsible for a series of problems: from the loss of valuable farmlands to residential and commercial development, to the increased costs of local government services, the spatial segregation of racial and ethnic groups, and air pollution and consequent global warming (e.g., Dreier *et al.*, 2001: 30-55; Newman & Kenworthy, 1989; Miller, 2008). These critics of sprawl cite among its causes the lack of appropriate governmental policies to curb sprawl, or the presence of misguided governmental policies that promote sprawl; they propose policy alternatives like "smart growth," "new urbanism," "urban containment," or creating centralized metropolitan governments as solutions (e.g., Katz, 1994; Orfield, 1997; Rusk, 1999).

On the other side of the debate are those who question the evidence the critics use to support the argument that sprawl causes all the above-cited problems (e.g., Bruegmann, 2005; Nivola, 1999). Some argue that sprawl has actually desirable outcomes (e.g., Barnes, 2000; Bruegmann, 2008; Easterbrook, 1999). Others argue that regardless of the problems it might create, sprawl is a natural outcome of individuals making rational decisions: Whenever they have the ability, people tend to move away from the congestion and blight of the cities toward the larger and well-maintained houses and yards in suburbs (primary examples of this approach are public choice theorists, as cited in Dreier *et al.*, 2001: 97). Another argument is that anti-sprawl policies are undesirable, because they distort the land markets in urban areas by making housing less affordable for large numbers of people (e.g., Bruegmann, 2005: 204).

Implicit in the conceptualizations on both sides of the debate are the notions that urban sprawl can be defined universally and that its linear causes and consequences can be identified. While the critics aim to find out the definitive causes of sprawl and devise policy solutions accordingly, their opponents make the assumptions that universal laws guide human behavior (such as that all individuals make utility maximizing rational choices) and that any restrictive governmental intervention will have undesirable consequences. Both sides acknowledge that the phenomenon of sprawl is more complex than their own theories. However, because they both theorize in the deductive-nomological mode and make linear assumptions about human behaviors, economy, and policy, they fall short of offering conceptualizations that would help us understand the complexity of sprawl. My argument is that complexity theory, with its epistemological bases and methodological tools, is better equipped to do that.

Complexity theory encourages us to think in systemic and dynamic terms. Complexity theory counters assumptions such as that policy problems can be defined categorically and definitively (e.g., "Urban sprawl is…."), that their linear causes can be discovered (e.g., "Wrong policies cause sprawl."), and that public policy tools can be deployed to solve these problems in a linear fashion (e.g., urban containment policies or laissez faire policy approaches would solve the problem). As I explain briefly later in the paper, and discuss more extensively elsewhere (Morçöl, 2012), complexity theory suggests that social problems defy categorical universal definitions and linear solutions. Researchers should instead understand them contextually. Also, public policies should be understood as actions of governmental actors that contribute, together with the actions of other actors, to the emergence the macro properties of complex urban systems, such as sprawl.

In this paper I present an outline of how complexity theory can be applied to studying urban sprawl and discuss the conceptual and methodological problems that need to be addressed in doing so. In the following section, I discuss the definitions and causes sprawl that are cited in the literature. I illustrate the problems with these definitions and attributed causes in a separate section. In the final section, I present the complexity framework in understanding sprawl and discuss its implications for public policy.

Is There A Theory Of Sprawl?

There is no universally accepted definition of urban sprawl; nor are there universally accepted causes of it. As Bruegmann (2001) notes, actually the term "sprawl" is used in the literature in an "imprecise and highly evocative" manner (16087). The most general definition of urban sprawl is that it is a form of the expansion of urban lands. Then, one might ask, is it merely another name for the natural growth pattern of urban areas? After all, as human populations grew, so did the land they occupied in throughout history (Mumford, 1961: 482-483).

There are more specific, but still problematic, definitions of sprawl in the literature. Bruegmann (2005) defines urban sprawl as "low-density, scattered, urban development without systematic large-scale or regional public land-use planning" (18). There are problems with the universal applicability of this definition. First of all, density is a geographically and historically relative term. "Low-density" by European standards may be relatively high-density by American standards (Bruegmann, 2001). Also, a settlement that is low-density at one time may be become denser at a later time. Density is also historically relative term: The ancient Egyptian and Mesopotamian cities were not as densely populated compared to today's New York, Paris, or London, for example (Mumford, 196: 482-483).

The problem with citing the lack of "regional public land-use planning" as part of the definition of urban sprawl, as Bruegmann does, is that there is a wide range of policies and planning tools applied in urban areas (regulating land use, taxation policies, etc.) and each may have different consequences.

Bruegmann's definition of sprawl as "scattered development" seems to follow Mumford's (1961) observation that suburbs in America have developed without a form (he calls this "formless urban exudation," p. 505). This was a result of the mass production of uniformly styled houses since the early 20th century (p. 486). The problem with defining sprawl in terms of "scattered developments," or as "formless exudation," is that it does not help us understand the problems experienced in the high-density and well-organized "new downtowns" in suburbia (Lewis, 1996: 2). Some of these problems are similar to those experienced in "scattered" suburbia, such a long commuting times and consequent air pollution, as I will illustrate with the examples of Portland and British cities in this paper.

There are also numerical definitions of sprawl. Couch and his colleagues (2007) define it in terms of the "urban population density gradient." Population density declines as one moves away from urban centers to suburbs and exurbs. In their definition, an urban area would be considered sprawled if the gradient is less steep than the "normal gradient" (5-6). The problem with this definition is that, as the authors recognize, what is a "normal" gradient is relative. What is "normal," for example, is different between America and Europe and between northern and southern parts of Europe (16).

Urban sprawl is also defined in terms of the differences between cities and suburbs in their rates of population and economic growth. Batty (2007) notes that in the early 19th century suburban fringes began growing faster than urban centers in America and Britain; he points to this relatively faster rate of growth of suburbs as the signature of sprawl (386). This observation needs to be qualified, however: There are variations among the rates of growth not only between cities and suburbs, but also among various cities and among various suburbs, as Batty also notes (387).

Although it is not possible to elicit a common and precise definition of sprawl from the above discussions, a general working definition can be proposed: A sprawled urban area is one where outer settlements have relatively lower population densities and grow faster than core urban areas; these areas may take on various spatial forms. Government policies can influence, but do not necessarily determine, these forms.

Different authors mention different causes of sprawl. Nivola (1999), for example, cites natural factors (e.g., availability of large open spaces), demographic factors (higher rates of population growth in the United States), social factors (e.g., destabilizing effects of urban violence), cultural factors (e.g., tendency of ethnic groups to separate themselves from others geographically), and technology (e.g., wide availability of cars), as the causes of sprawl in American urban areas (4-11). Bruegmann (2005) identifies a few more causes that are commonly cited in the literature: from the American individualistic culture that encourages people to live in their separate lands, to the unfettered capitalism and lack of governmental regulations in the United States (96-112). Others find various governmental policies responsible for the sprawl in the US urban areas: from the federal homeowner subsidies, to the federal tax code, the building of the Interstate Highway system, to the local zoning ordinances that segregated urban functions into geographically distinct areas (Dreier *et al.*, 2001; Rusk, 1999).

Bruegmann (2005) argues that affluence and democracy are the main causes of suburbanization. These two conditions allow individuals to be more mobile and give them the choice to move to places where they can have privacy (109-112). He admits, however, that these two cannot universally explain sprawl, because so many wealthy citizens of democratic countries choose to live in dense urban areas, such as Park Avenue, New York, and the Sixteenth Arrondissement in Paris.

One can find some commonalities in the causes of sprawl that are cited in the literature, but there is no definitive list. The lack of a common and precise definition of sprawl and the multiplicity of its explanations in the literature indicate that it is a complex phenomenon that poses conceptual and methodological challenges.

I further illustrate these challenges with the summaries of three cases in the next section: the metro areas of Atlanta and Portland in the United States and the national case of Britain. These cases are selected because the urban areas in the United States are generally considered the most sprawled in the world and British governments developed the most-well known anti-sprawl policies in the world. Atlanta and Portland are known as the examples of two opposite policy approaches to sprawl in the United States. These cases will illustrate that urban sprawl is actually more complex than its categorical characterizations and that it takes on different forms that emerge in the particular historical and political contexts of the cities and nations.

Case Summaries

Atlanta

The information in this section is based on the case studies conducted by Morçöl and his colleagues (2003) and Zimmermann and his colleagues (2003), unless noted otherwise. If there is a "poster child of sprawl," metro Atlanta can be considered a primary candidate for this title. It is called "the fastest-growing human settlement in history regarding land consumption" (Leinberger, 2008). The details of how this came about are not straightforward, nor are the current outcomes definitive.

The population of metro Atlanta increased from 1.7 million in 1970 to 5.7 million in 2010, while the population of the city declined from 495.000 in 1970 to 395.000 in 1990 and then steadily increased to 420.000 in 2010. The city's share of the metro area population declined from 29% in 1970 to 7% in 2010. The Atlanta metropolitan statistical area covered five counties in 1970; this has expended to 20 counties in 2000. The built environment of the metro area stretches about 100 miles from one end to the other. If sprawl is defined as "formless urban exudation" with low population density, Atlanta's suburbs with no, or little, discernible areas of concentration fit this definition. The population densities in the city and the metro area (3,160 per square mile and 1,800 per square mile respectively) are much lower compared to some major metro areas in the United States, such as New York and San Francisco (http://2010.census.gov/2010census/), but higher compared to others like Denver, Seattle, Minneapolis-St. Paul, and New Orleans (Jaret, 2002: 169).

Atlanta's rapid expansion can be attributed partly to the gradual shift of the manufacturing base and population in the United States from the Northeast and Midwest to the South and Southwest since the 1960s. The pro-growth poli-cies of the state government in Georgia and the "urban regime coalitions" that controlled the city's politics in Atlanta since the 1950s (Stone, 1989) can also be cited as reasons. Key aspects of these policies were to keep building high-ways stretching to farther geographic areas and to allow building new suburban subdivisions with little or no restrictions. Also, as the percentage of the black population increased in the city, large sections of the affluent and middle-class whites moved out and settled in suburbs, which contributed to the suburban expansion. These economic and demographic shifts created self-organizational dynamics: Suburban economic growth attracted more growth in suburbs, while the economic activities and population in the city declined until the early 1990s.

An important consequence of the rapid growth was the decline in the air qual-ity in the metro area. As more highways were built, metro Atlantans settled in distant locations, commuted longer distances, the traffic congestion intensified, and large amounts of pollutants were emitted to the atmosphere. In the late 1990s the US Environmental Protection Agency (EPA) declared the Atlanta region a serious violator of the National Ambient Air Quality Standards, and threatened to cut federal funds for further road building. Equally important, Atlanta's image would be tarnished because of the pollution and traffic problems and further economic growth could be stymied. Metro Atlanta's business leaders and the state government reacted to EPAs threat quickly. In 1999, with the support of the metro business leaders, the governor and the state general assembly cre-ated the Georgia Regional Transportation Authority (GRTA) and authorized it to review large construction projects, including transportation projects. The GRTA board initially drafted plans to expand the mass transit system throughout the region. However, when the governor lost the 2002 elections for unrelated rea-sons (i.e., the backlash of voters to the governor's leadership in replacing the state's flag, which had the "Confederate Battle Flag" as part of it, with a new flag in 2001) this created new dynamics. Under the new governor GRTA's mass tran-sit initiatives were slowed down and even halted, while the road construction continued.

Arguably the policy intervention to curb sprawl in the early 2000s failed, but the effects of this failure are not clear cut. On the one hand, suburbs continued to grow faster than the city (the city's share of the metro area population continued to decline from 10% in 2000 to 7% in 2010). On the other hand, the city's popula-tion increased, albeit slowly, between the 1990 and 2010, as noted above, and the city's neighborhoods have become denser and the occupancy rates in resi-dential and commercial units in the central areas increased (Leinberger, 2008).

Britain

The British "urban containment" policies that have been implemented in the last half a century represent a contrast to the pro-growth policies implemented in metro Atlanta in the same period. Champion's (2003) case study of the British policies show that these policies did have effects on the evolutions of the spatial and social forms of urban areas, but in nonlinear ways, with both intended and unintended consequences.

The British policies were rooted in the Greater London Plan of 1945, which imposed an urban development boundary and an area beyond this boundary was designated as the "green belt," where urban development would be controlled tightly. The excess population growth in London would then be directed to the concentrated "new towns" outside this belt (Milward, 2006). Successive policies of the national governments in the following decades expanded this initial London model to other cities in Britain.

The British containment policies led to the relatively high densities in urban cores, as well as in self-contained distant settlements, as intended (Champion, 2003: 14, 64, 74). There have been unintended consequences as well. As the outmigration from cities intensified, more people began to commute longer distances, which increased the traffic congestion and air pollution in London and other cities. The nation's economic base moved out of cities at rates faster than population growths in outer regions (48). Champion also observes that two kinds of separation happened in urban areas: the separation of residential and commercial functions and the separation of affluent "shire" counties and deprived cities (75). Champion also argues that the policies increased land and housing prices (15).

Portland

The British containment policies have been emulated in Japan, Canada, and the United States (Milward, 2006). Over a hundred US cities have adopted similar policies since the 1950s (Nelson *et al.*, 2008: 9). Portland, Oregon, is the most prominent of these cities and its experience has been debated most intensely in the literature. If Atlanta is the poster child of sprawl, then Portland is the poster child of the systematic efforts to curb sprawl.

Oregon adopted a state-wide urban growth and management policy with its 1973 Conservation and Development Act (Bruegmann, 2005: 204). This state law was implemented on the largest scale in Portland. Two key components of the implementation in Portland were setting an "urban growth boundary" and shifting the priorities in transportation from constructing new highways to making investments in public transit (Bruegmann, 2005: 204). A metropolitan area-wide special district, known as Metro, was charged with drawing the urban growth boundary initially; over time Metro became an elected body and granted wider authority (Bruegmann, 2005: 205-206).

Have policies in Portland been successful? The answer depends on who answers the question and which set of statistics they use to support their arguments.

General population statistics indicate that the Portland metro area has expanded geographically without interruptions since the 1950s. The population density declined until the early 1980s, when the "urban growth boundary" policy went into effect, and since then it has increased slightly (Abbott, 2002: 221). It is not clear if this increase was a result of the containment policy, because in the 1980s the population densities of some other US metro areas either stabilized or increased as well, while in some others it continued to decline (Bruegmann, 2005: 62-63). The city's population density continued to decline in the 1980s, but it began to increase in the 1990s (Bruegmann, 2005: 62-63, 205-206). Even in the 1990s, population grew relatively slowly in the city, while the populations of metro Portland's outlying areas (Vancouver, Washington and other nearby towns) grew faster (Bruegmann, 2005: 209). Meanwhile, the share of the central business district employment in the metro area increased significantly in the 1980s, contrary to the trend in other US cities in the same period (Lewis, 1996: 169).

In the area of transportation, again, there are mixed and even conflicting results. The car usage rates in the city and the metro area have increased since the 1970s and the market share of the public transit has declined (Bruegmann, 2005). Because of the population shifts to nearby towns, the number of daily car commutes to Portland increased in these decades as well (209). Lewis (1996) provides a set of statistics that depict a different picture: Bus, rail, bike, and walking accounted for 40% of the commutes to Portland's central business district and in Portland metropolitan statistical area public transportation accounted for 9.2% of work trips in the mid-1990s (195). Bruegmann, on the other hand, notes that the percentage of people using public transit declined to less than 1% by the mid-2000s (212).

Bruegmann compares Portland to Phoenix, Las Vegas, and Houston and notes that in Portland housing values have risen much faster (210). Lewis (1996) compares Portland to San Diego, Sacramento, and Seattle and says that Portland is three times more affordable than the other three (181).

Lessons Learned From The Cases

The above case summaries show that urban sprawl does not have a universal form. The effects of public policies on urban expansion are not uniform or linear either. There are some commonalities among different cases, however. Metro Atlanta sprawled in a way that fits Mumford's description of "formless urban exudation." In Britain urban areas expanded, but they did so in denser pockets around major cities. Similar to the process in Atlanta, in Britain the populations of outer settlements grew faster than cities, economic activities moved outside cities, socio-spatial separation happened between cities and outer settlements, and traffic congestion and air pollution increased.

The Atlanta example may be cited to support the argument that laissez faire pro-growth policies of the business leaders/elites and local and state governments encourage demographic and geographic expansion. The experiences in Britain and Portland show that their containment policies did have some effects, but there are similarities between them and Atlanta in some important respects as well. For example, the degrees of dependence on cars in daily commutes have increased in all the metro areas studied. The Portland experience shows that the containment policies did not have clearly discernible effects on population growth, employment patterns, or the utilization of the modes of transportation. The population densities in the cities and surrounding areas in Atlanta and Portland decreased and increased over time in no direct correlation with the respective policy changes. Sprawl is a dynamic phenomenon

A Complexity Theory Of Urban Sprawl

The case summaries above illustrate that sprawl is a complex phenomenon. The lack of precise definition of sprawl and the inability of researchers to identify definite and linear causes are due to this complexity. The summaries also indicate that in each case multiple factors and actors played roles in the expansions of urban areas in somewhat different forms with some similarities. As Bruegmann (2005) points out, sprawl is generated by "innumerable forces, always acting on each other in complex and unpredictable ways" (112). This complexity defies the efforts to define sprawl universally and finding universal causes of it. It also defies the efforts to fix social problems such as sprawl with "policy solutions."

Complexity theory offers an alternative conceptual understanding of these systems and the roles policies play in them and provides a set of methodological tools to investigate these systems. Urban sprawl can be better investigated and understood within this alternative view.

Applying Complexity Theory Concepts And Methods

Complexity theory suggests that urban spatial forms, such as sprawl, are emergent properties of complex urban systems. These systems are self-organizational in the sense that the patterns of urban social and spatial processes cannot be dictated or prevented by external interventions, such as government policies. They are dynamic in the sense that urban spatial and social forms change over time, as urban systems and urban actors interact with natural, economic, and cultural systems. Because the relationships among the actors of urban systems are nonlinear, the emergent properties of these systems (e.g., how and to what extent urban areas expand) cannot be defined categorically, nor can their causes be identified with certainty.

Each urban actor (politicians, businesses, home owners etc.) has complex belief systems and motivations, which makes predicting their behaviors and their interactions nonlinear (Haag, 2002: 14). These complex belief systems and non-linear interactions limit researchers' ability to single out a simple and universal explanation of sprawl and devise policies accordingly. As Batty (2007) puts it, complex systems defy the "conventional view of science," which seeks simplicity. Within the complexity view: "[P]redictability and certainty, the traditional hallmarks of science, indeed the traditional hallmarks of urban theory a generation or more ago, are in question" (515). The unpredictability and uncertainty in knowledge processes limit researchers' abilities to make generalizations.

This is why, from the perspective of complexity theory, it is not possible to define sprawl universally or identify a set of universal causes of sprawl in the deductive-nomological sense of the term. Neither the case studies summarized earlier in this paper, nor others described in other sources suggest that there is a universal form sprawl or a set of causes attributable to it. Instead, complex urban processes and the emergent properties of urban systems, such as sprawl, should be studied within their contexts. Researchers can factor the similarities observed among different urban areas into their models, but they should also aim to understand the individual characteristics of each case.

This contextual investigation involves two theoretical and empirical problems/questions. First, how do the relationships of an urban system with other systems (e.g., national economy; ecological environment) affect urban sprawl? This is the problem of coevolution of systems. Second, how do macro-level (systemic) properties, such as urban sprawl, emerge from micro-level interactions (i.e., the interactions among individual and collective actors)? This is the "micro–macro problem" or the "agency–structure problem."[1]

As early as in the 1960s, scholars like Forrester (1969) used system dynamics modeling to investigate the effects of the interactions among social systems and natural systems on urban expansions. Haase and Seppelt's (2010) work is a more recent example. As useful as these macro-level analyses are, to understand complex urban processes better, researchers should investigate the micro–macro transformations in urban systems. As Occelli (2006) observes, beginning with the 1980s, urban researchers have made some advances in doing so.

To understand the importance of these advances, we need to define the micro–macro problem first. Coleman (1986) defines it as the problem of understanding how "individual preferences become collective choices" (1321). Its philosophical roots go back to Hobbes, Adam Smith, Locke, Rousseau, and Mill. Talcott Parsons posed it as the central problem of sociology in the 1930s. Since then scholars from various theoretical perspectives have adopted it as their core intellectual

---

1. Actually this question represents only one aspect of the micro–macro problem. The problem includes the following questions as well: Are the properties of the emergent structures irreducible to those of agents? (How) do emergent properties influence the actions of agents? These questions and their implications are discussed extensively elsewhere (Morçöl, 2012).

E:CO Vol. 14 No. 4 2012 pp. 1-16

problem (e.g., Jessop, 1991; Ostrom, 1990, 2005). Some complexity researchers have done so as well. Haag's (2002) formulation of the problem of the complexity of urban sprawl in micro–macro terms is of particular relevance here:

> [U]rban sprawl is the result of an interlocked process of spatial interactions where different agents (households, accommodation agencies, employees, firm, etc.) with different, partly inconsistent interests, are involved. The multiple decisions of the different agents result in migration flows of…people, changes in commuter flows and…in a redistribution of workplaces in an extending spatial region. (13)

Then, how can we know how the decisions by multiple urban agents and their interlocked spatial interactions result in migration flows, commuter flows, the redistribution of workplaces, affect the extension of urban areas? The primary methodological tool of complexity researchers in investigating micro–macro problems is agent-based simulations, or multi-agent simulations. Batty's (2007: 403-415) cellular automata simulations of urban growth processes, Hass's (2002: 34-42) simulations of the population dynamics in Stuttgart, and Lüdeke and his colleagues' (2007) comparative simulations of several urban areas in Europe are good examples of the applications of these methods to urban problems. Hass's and Lüdeke and his colleagues' simulations are particularly illustrative of the observation I made above that the causes and forms of sprawl cannot be generalized to all countries or urban areas; instead cases should be investigated within their contexts.

Implications For Policymakers And Planners

What are the implications of complexity theory for those policymakers, policy analysts, and planners who are concerned about urban problems in general, and sprawl in particular? The most obvious implication is that they should recognize the limits of their ability to shape urban processes. This is particularly because complexity theory challenges the presumed causal relationship between policy decisions and actions on one side and "policy outcomes" on the other (Salzano, 2008: 186). Such a causal relationship cannot be established, as I pointed out earlier, because of the nonlinearity of the relationships among multiple policy actors and the self-organizational nature of complex systems, such as urban systems. Instead, policymakers and planners should recognize the emergent nature of urban systems.

This complexity theory insight into complex urban systems has precedents in the 1960s. Jacobs (1993, first published in 1961) argued against the massive "urban renewal" projects of her time and for letting neighbors of "blighted areas" organize themselves to renew their neighborhoods. Forrester (1969) observed that urban systems are self-organizational and that they resist interventions by governments: They are uncontrollable and unplannable.

The case studies summarized above demonstrate that anti-sprawl policies, such as greenbelts, cannot control urban expansions in the exact ways that were intended by policymakers. British cities and Portland expanded geographically and the traffic congestion increased in their respective metropolitan areas, despite the policies. The Atlanta experience demonstrates that the sprawl process may slow down despite the policy failures, like the failure of GRTA. These cases do not indicate, however, that there is no relationship between policies and outcomes; they rather illustrate that the relationships are nonlinear.

Then what should policymakers and planners do? A generic answer to this question is that they should adopt the concepts and methodological tools of complexity theory in their policymaking and planning activities. I must stress, however, that complexity theorists have not offered a coherent and unified set of concepts and tools for policymakers and planners yet. Theirs is a work in progress.

The most intuitively appealing concept of complexity theory is self-organization. In Portugali's (2000) view urban planning should be self-organizational in the sense that planners and policymakers should adopt general planning principles and let urban agents plan for themselves, instead of imposing specific land-use plans on them. In this bottom-up approach, urban actors would provide planning ideas and planning parameters would emerge from them. The role of planners would be to provide information and technical expertise to urban actors and let them interpret the information and make and update their plans.

Self-organization is not a uniform process, nor is it a panacea, however. As Buijs and his colleagues' (2009) case study of the Randstad Holland metropolitan region in the Netherlands demonstrate, self-organizational policymaking and planning is a highly complex process. To curb the urban sprawl in the region, the Dutch central government created mechanisms of collaboration in land-use planning among the local governments in this region. Buijs and his colleagues' study shows that self-organization may take on multiple forms and the broader economic and political context (e.g., the framework set by the central government) matter in enabling self-organization. Many decades of studies by Elinor Ostrom and her colleagues also show that self-organization is not a uniform process and that certain conditions should be created for self-organizational policy/planning processes to be effective (see Ostrom, 2005: 244-245).

Although complexity theory encourages us to think of policy and planning processes in nonlinear and self-organizational terms, this does not mean that effective policy or planning mechanisms cannot be devised. Complex policymaking and planning processes can be effective if governmental and private actors together create and maintain "robust but adaptive governance systems" (Bankes, 2008) or "resilient and adaptive governance systems," which would "withstand shocks and surprises, absorb extreme stresses, and maintain [their] core functions, through perhaps in…altered form[s]" (Innes & Booher, 2010: 205). If the expansion of urban land and decreasing population density in metropolitan areas are problems, for example, governmental actors and non-governmental actors together can affect land-use patterns and densities.

How can complex policy systems be made "robust" or "resilient" is a major area of research. In their studies Elinor Ostrom and her colleagues identified a set of principles of "designing robust social-ecological systems" (see Ostrom, 2005: 258-279). Their studies provide guidelines to policymakers and planners in areas such as how to make collective-choice arrangements, establish accountability and sanctions, and resolve conflicts. Miller and Page (2007) point out that complexity researchers need to do more research to understand better the robustness and adaptability of complex systems in the face of the changes in their environments and the changes in their actors and their relationships (236-237).

Current and future studies by complexity researchers in these areas will help us better understand how urban spatial forms emerge in general and how urban sprawl occurs in particular. They will also help us develop better guidelines to enable the creation of robust and adaptable urban systems to make our communities more livable and less harmful to their natural environments.

References

Abbott, C. (2002). "Planning a sustainable city: The promise and performance of Portland's urban growth boundary," in G.D. Squires (ed.), *Urban Sprawl: Causes, Consequences & Policy Responses*, ISBN 9780877667094, pp. 207-235.

Barnes, F. (2000, May 22). "Suburban beauty: Why sprawl works," *Weekly Standard*, ISSN 1083-3013, pp. 27-30.

Batty, M. (2007). *Cities and Complexity: Understanding Cities with Cellular Automata, Agent-Based Models, and Fractals*, ISBN 9780262025836.

Bruegmann, R. (2001). "Urban sprawl," *International Encyclopedia of the Social & Behavioral Sciences*, ISBN 9780080430768, pp. 16087-16092.

Bruegmann, R. (2005). *Sprawl: A Compact History*, ISBN 9780226076911.

Bruegmann, R. (2008). "Concerns about urban sprawl are class-based objections to middle-class developments," in D.A. Miller (ed.), *Urban Sprawl*, ISBN 9780737739664, pp. 59-67.

Buijs, Jean-Marie, Van der Bol, Nancy, Teisman, Geert R., and Byrne, David. (2009). "Metropolitan regions as self-organizing systems," In G. Teisman, A. van Buuren, and L. Gerrits (eds.), *Managing Complex Governance Systems: Dynamics, Self-Organization and Coevolution in Public Investments*, ISBN 9780415459730, pp. 97-115.

Coleman, J.S. (1986). "Social theory, social research, and a theory of action," *The American Journal of Sociology*, ISSN 0002-9602, 91(6): 1309-1335.

Couch, C., Leontidou, L., and Arnstberg, K-O. (2007). "Introduction: definitions, theories, and methods of comparative analysis," in C. Couch, L. Leontidou, and K-O. Arnstberg (eds.), *Urban Sprawl in Europe: Landscapes, Land-Use Change & Policy*, ISBN 9781405139175, pp. 3-38.

Dreier, P., Mollenkopf, J., and Swanstrom, T. (2001). *Place Matters: Metropolitics for the Twenty-First Century*, ISBN 9780700611348.

Durning, D.W. and Brown, S.R. (2007). "Q methodology in decision making," in G. Morçöl (ed.), *Handbook of Decision Making*, ISBN 9781574445480, pp. 537-564.

Easterbrook, G. (1999, March 15). "Suburban myth: The case for sprawl," *New Republic*, ISSN 0028-6583, 18-21.

Epstein, J.M. and Axtell, R. (1996). *Growing Artificial Societies: Social Science from the Bottom up*, ISBN 9780262550253.

Forrester, J.W. (1969). *Urban Dynamics*, ISBN 9781883823399.

Gainsborough, J. (2001). *Fenced off: The Suburbanization of American Politics*, ISBN 9780878408313.

Innes, J.E. and Booher, D.E. (2010). *Planning with Complexity: An Introduction to Collaborative Rationality for Public Policy*, ISBN 9780415779326.

Jacobs, J. (1993). *The Death and Life of Great American Cities*, ISBN 9780679600473.

Jaret, C. (2002). "Suburban expansion in Atlanta: The 'city without limits' faces some," in G.D. Squires (ed.), *Urban Sprawl: Causes, Consequences & Policy Responses*, ISBN 9780877667094, pp. 165-205.

Jessop, B. (1991). *State Theory: Putting the Capitalist State in its Place*, ISBN 9780271007458.

Kane, M. and Trochim, W.M.K. (2007). *Concept Mapping for Planning and Evaluation*, ISBN 9781412940283.

Katz. P. (1994). *The New Urbanism: Toward an Architecture of Community*, ISBN 9780070338890.

Kelly, G.A. (1955). *A Theory of Personality: The Psychology of Personal Constructs*, ISBN 9780393001525.

Leinberger, C.B. (2008, November 5). "Sprawl to meet its limit in Atlanta," *Atlanta Journal and Constitution*, http://www.ajc.com/services/content/opinion/stories/2008/11/05/leinbergered.html (no longer available).

Lewis: G. (1996). *Shaping Suburbia: How Political Institutions Organize Urban Development*, ISBN 9780822955955.

Lüdeke, M., Reckien, D., and Petschel-Held, G. (2007). "Modeling urban sprawl: Actors and mathematics," in C. Couch, L. Leontidou, and K-O. Arnstberg (eds.), *Urban Sprawl in Europe: Landscapes, Land-Use Change & Policy*, ISBN 9781405139175, pp. 183-216.

Miller, D.A. (ed.) (2008). *Urban Sprawl*, ISBN 9780737739664.

Miller, J.H. and Page, S.E. (2007). *Complex Adaptive Systems: An Introduction to Computational Models of Social Life*, ISBN 9780691127026.

Milward, H. (2006). "Urban containment strategies: A case-study appraisal of plans and policies in Japanese, British, and Canadian cities," Land Use Policy, ISSN 0264-8377, 23: 473-485.

Morçöl, G. (2012). *A Complexity Theory for Public Policy*, ISBN 9780415518277.

Morçöl, G., Zimmermann, U., and Stich, B. (2003). "The Georgia Regional Transportation Authority: A smart growth machine?" *Politics and Policy*, ISSN 1747-1346, 31(3): 488-511.

Mumford, L. (1961). *The City in History: Its Origins, Its Transformations, and its Prospects*, ISBN 9780156180351.

Nelson, A.C., Dawkins, C.J., and Sanchez, T.W. (2008). *The Social Impacts of Urban Containment*, ISBN 9780754670087.

Newman P. and Kenworthy, J. (1989). *Cities and Automobile Dependence: An International Sourcebook*, ISBN 9780566070402.

Nivola, S. (1999). *Laws of the Landscape: How Policies Shape Cities in Europe and America*, ISBN 9780815760818.

Occelli, S. (2006, May). "Technological convergence vs. knowledge integration," paper presented a Les Journées Annuelles Transdisciplinaires de réflexion au Moulin d'Andé, Colloques AFSCET, May 13-14, 2006, http://www.afscet.asso.fr/soAnde06.pdf.

Occelli, S. and Staricco, L. (2006). "Cognitive stances in urban mobility: A simulation experiment," *Language and Cognitive Processes*, ISSN 0169-0965, 7(Suppl. 1): S72-S74

Orfield, M. (1997). *Metropolitics: A Regional Agenda for Community and Stability*, ISBN 9780815766391.

Ostrom, E. (1990). *Governing the Commons: The Evolution of Institutions for Collective Action*. ISBN 9780521405997.

Ostrom, E. (2005). *Understanding Institutional Diversity*, ISBN 9780691122076.

Portugali, J. (2000). *Self-Organization and the City*, ISBN 9783642084812.

Rusk, D. (1999). *Inside Game/Outside Game: Winning Strategies for Saving Urban America*, ISBN 9780815776512.

Salzano, M. (2008). "Economic policy hints from heterogeneous agent-based simulations," in K. Richardson, L. Dennard, and G. Morçöl (eds.), Complexity and Policy Analysis: Tools and Methods for Designing Robust Policies in a Complex World, ISBN 9780981703220, pp. 167-196.

Schwarz, N., Haase, D., and Seppelt, R. (2010). "Omnipresent sprawl? A review of urban simulation models with respect to urban shrinkage," *Environment and Planning B: Planning and Design*, ISSN 0265-8135, 37: 265-283.

Stone, C.N. (1989). *Regime Politics: Governing Atlanta, 1946-1988*, ISBN 9780700604166.

Zimmermann, U., Morçöl, G., and Stich, B. (2003). "From sprawl to smart growth: The case of Atlanta," in M.J. Lindstrom and H. Bartling (eds.), *Suburban Sprawl: Culture, Theory, and Politics*, ISBN 9780742525801, pp. 275-288.

Göktuğ Morçöl is an associate professor of public administration and policy in the School of Public Affairs at Penn State Harrisburg. His research interests are complexity theory, metropolitan governance, business improvement districts, and policy analysis and evaluation methodology. He has authored, edited, and coedited the following books: *A Complexity Theory for Public Policy* (Routledge, 2012), *Complexity and Policy Analysis* (ISCE Publishing, 2008), *Business Improvement Districts* (CRC Press, 2008), *Handbook of Decision Making* (CRC Press, 2007), *A New Mind for Policy Analysis* (Praeger, 2002), *New Sciences for Public Administration and Policy* (Chatelaine Press, 2000). His articles have appeared in *Public Administration Review, Administrative Theory & Praxis, Policy Sciences, Public Administration Quarterly, Politics and Policy, International Journal of Public Administration, Emergence,* and other journals.

Applied

Management As System Synchronization: The Case Of The Dutch A2 Passageway Maastricht Project

Stefan Verweij
Department of Public Administration, Erasmus University Rotterdam, NED

On the one hand, the importance of flexibility and adaptiveness in the design and management of human activity systems to deal with complexity is stressed. On the other hand, existing frameworks of procedures, practices and rules often require strict planning, design and implementation. This raises the question how flexibility and adaptiveness comply with these existing frameworks to arrive at effective and efficient project realization. A grounded analysis of the Dutch infrastructure project A2 Passageway Maastricht, instigated by the question how the influence of the management system on the provisional outcomes of the project can be explained, found that it involves system synchronization: combined system fragmentation and integration.

Introduction

In 2009 the Organization for Economic Cooperation and Development (OECD) published a report in which the important role of complexity science for public policy and management is stressed. Other recent publications and special issues in the fields of Public Administration, Public Policy and Public Management are exemplary of the need for complexity sciences in the public realm (Dennard *et al.*, 2008; Meek, 2010; Rhodes *et al.*, 2011; Teisman & Klijn, 2008; Teisman *et al.*, 2009b). Whereas conventional management assumes a stable hence predictable world that can be managed and controlled by developing and implementing plans, visions and strategies (Stacey, 1992), these recent publications stress the need for more complexity-informed design and management of human activity systems for the effective and efficient realization of projects. It is stressed that policy making and management have to match the complexity encountered. For instance, literature on the management of complex systems stresses the need for flexible, adaptive and contextual management strategies (Edelenbos *et al.*, 2009).

Although there is plenty of evidence for this need, the call for this special issue of *Emergence: Complexity & Organization* rightly identified the challenge of how these insights from complexity science are to comply with existing public procedures, practices and rules that often involve strict planning, design and implementation. It is thus of interest to learn how complex system management interacts with this existing institutional context and how this influences the effective and efficient realization of projects. The aim of this article is therefore to show how management practices take shape in practice and to explain how these influence provisional outcomes in projects. This is done by presenting a case study of the management of the Dutch coupled infrastructure-area development project A2 Passageway Maastricht.

This article is structured as follows. The research framework is presented in the sections "Research Framework" and "Methods". More specifically, "Researching Social Complexity" briefly discusses the nature of social complexity and its implications for researching it (cf. Gerrits & Verweij, forthcoming). Importantly, it argues that a grounded research approach is required. The "Analytical Framework" section presents the subsequent conceptual framework to analyze this social complexity. The next section, "Methods", discusses the methods used in data collection and analysis that follow from the perspective outlined in "Researching Social Complexity" and "Analytical Framework". The "A2 Passageway Maastricht Infrastructure Project" is introduced next, based on an analysis of documents and webpages. The empirical results of the analysis of interviews about management practices are presented in "Interview Analysis". In the penultimate section, "Managing Complexity Through System Synchronization", these results are interpreted and linked with theory. The article concludes with a discussion of the findings resulting from the grounded analysis.

Research Framework

Researching Social Complexity

Complexity is recursive to the extent that an ever closer look into it reveals ever more details that mirror the whole but not in the way that the whole can be fully known and understood. As explained by Cilliers (2001), any model of complexity is incomplete by definition. This is inherent to the concept of model. Moreover, complexity in the social realm is dynamic: "what constitutes and limits a system is relative to the agents' and observers' locality (…). Therefore, complexity arises not only from the constituent elements of a system, but also from the fact that this constitution is dynamic itself" (Gerrits, 2008: 20-21).

The fundamental statement about complexity is that the world is composed of open systems which are nested and have nested within them other open systems (Byrne, 2005). The basic elements of social systems are reflexive actors: people who respond, anticipate, plan, think, forecast etcetera (Teisman *et al.*, 2009a). Social systems are bounded by interactions between these actors that are more or less sustainable. These interactions are reproduced through communication (Moeller, 2006) and motivated by certain aspects. Although bounded, systems interact with other open systems in their environment (Cilliers, 2001); hence they are contingent. They are situated in a particular context that brings about systemic changes since environmental influences can become part of the system's structure. The process by which this happens is emergence; reality consists of emergently-structured open systems (Reed & Harvey, 1992). However, emergence does not indicate a discrete entity or measurable phenomenon (Elder-Vass, 2005). Rather, it serves as a heuristic devise to think about the nature of causation. In addition, emergence points to the fact that predictive capacity is limited. These axiomatic statements imply that reality is essentially non-compressible (complexity is recursive) though we necessarily compress it, that reality is essentially non-decomposable (emergence) though we necessarily decompose it, that reality is contingent (nestedness of systems) and therefore explanation is local in time and place, and that reality is time-asymmetric (emergence) and hence predictive capacity limited (Gerrits & Verweij, forthcoming). This perspective has profound implications for researching public policy and management and the inferences that can be made (cf. Boulton, 2010; Waldrop, 1992: 331-334).

It resonates with critical realism (Byrne, 1998; Reed and Harvey, 1992) which is often referred to as the way forward between empiricism and interpretive science (Wuisman, 2005). Interpretation and construction of meaning play a pivotal role in understanding real complexity and causality (Gerrits, 2008). It implies the convergence of fact and value (Fischer, 1998) in the analysis of complex causation and acknowledges the locality of knowledge. Explanation of complex reality is thus temporal in time and local in place (Byrne, 2005; Gerrits, 2008) and based upon the boundary judgments (which involve compressions and decompositions) of agents involved in the systems (Buijs *et al.*, 2009; Gerrits, 2008).

These boundary judgments steer human activities in social systems (Buijs *et al.*, 2009; Teisman *et al.*, 2009a): through experiences agents construct models of reality upon which they act, so-called first order boundary judgments. These constructions are 'tested against' the reality that agents encounter and may be modified to present a better model of reality. In the words of Waldrop (1992: 253, original emphasis): "players operate in a world of *in*duction. They try to fill in the gaps on the fly by forming hypotheses, by making analogies, by drawing from past experience, by using heuristic rules of thumb. Whatever works—even if they don't understand why". And when people behave according to their perspective on the workings of the system they are part of, it becomes real as a consequence (i.e., Thomas Theorem).

Checkland (1981) and others (e.g., Uprichard & Byrne, 2006; Wagenaar & Cook, 2003) therefore argue that it is only through the eyes of people working in complex systems that a researcher can learn about their constitution and operations (cf. Wuisman, 2005). These people are in fact proverbial spies that help the researcher disclosing and understanding perceptions, actions and responses (cf. Buijs *et al.*, 2009). They are pivotal in two ways: they guide the researcher in generating temporal system boundaries, so-called second order boundary judgments (Buijs *et al.*, 2009), and they guide the researcher in generating preliminary causal relationships (Byrne, 2011).

Analytical Framework

The perspective outlined above requires a conceptual framework aimed at uncovering managers' models of reality for understanding the constitution—i.e., generating temporal boundary judgments—and operations—i.e., generating preliminary causal relationships—of social systems, and thus implies that these models are researched in a grounded manner (see also next section) by 'following the managers' (Checkland, 1981).

In this article the human activity system concerns the management system of the A2 Passageway Maastricht infrastructure project. This open system consists of reflexive agents—the managers—and interacts and coevolves with its environment (Gerrits, 2008). The boundaries of this management system are constituted by the researcher (i.e., second order) through the managers' accounts (i.e., first order) of memberships, interactions and goals of the system. These are considered as sensitizing concepts (Bowen, 2006). As Meadows (2008: 16) explains, membership and interactions can be observed relatively easy, but the third aspect of goals is "the least obvious part of the system [though it] is often the most crucial determinant of the system's behavior" due to i.a. the nested nature of systems. It can be argued that, since managers are part of organizations as social systems, organizational goals provide managers with motivations to act in a certain way (although their actions can also be based on individual motives) (cf. Mingers, 2003; Moeller, 2006; Seidl, 2005). More specifically, "to assess the rightness or feasibility of a particular action, practitioners employ criteria, standards and warrants that are part of a particular community or communities" (Wagenaar & Cook, 2003: 154). Hence, the goals of the project management system are inferred from the motivations of managers. The empirical constitution of the system is the subject of the section "Constituting Boundaries".

The constitution of systems is analytically distinct from the operations of the system, which are understood through the managers' accounts of actions. As explained by Uprichard and Byrne (2006: 668, original emphasis), "we want to know about what people say has happened, how they think things will develop, *and* how they will act in relation to those developments". These sensitizing elements (cf. Bowen, 2006) form the model of the actions of managers (i.e., first order) which the researcher infers from the data (i.e., second order).

The constitution and operations of systems are analytically distinct, but connected: the actions described by respondents are shaped by their motivations. Although shaped by motivations, actions are triggered by "everyday situations that because of some external or internal shift of events have suddenly become indeterminate and urgent (…) and therefore require some intervention by the actor" (Wagenaar & Cook, 2003: 150). The empirical operations of the project management system are the subject of the section titled "Operations Of The Management Systems".

It is important to note that, in order to understand the workings of a system, managers' accounts of actions, motivations and situations need to connect in a causal story of an empirical event (cf. Wuisman, 2005) upon which the research-er can generate provisional mini-theories (Weiss, 1994). A simplified version of the framework is depicted as Figure 1.

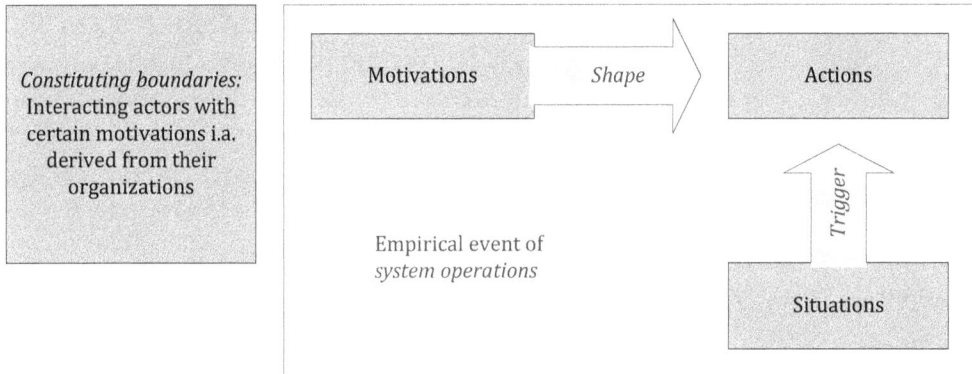

Figure 1 *Analytical Framework*

Methods

As explained, the complexity perspective and subsequent analytical frame-work described above imply, in line with Checkland's (1981) reasoning and that of others (Buijs *et al.*, 2009; Uprichard & Byrne, 2006; Wagenaar & Cook, 2003), a grounded and qualitative methodological framework using sen-sitizing concepts. For this study both document and interview data were col-lected and analyzed.

A total of 19 interviews were conducted. Details are provided in the appendix. Interviews were conducted in an open qualitative manner following the guide-lines and procedures as described by Weiss (1994). More specifically, interviews generally commenced with the open question 'can you tell me something about your role in the project' and from that point proceeded by picking up as many markers—generally defined as "a passing reference made by a respondent to an important event or feeling state" (Weiss, 1994: 77)—as possible. The focus was on the sensitizing concepts (membership, interactions, goals, motivations, actions and situations), but these were deliberately not operationalized (Schwartz-Shea & Yanow, 2012). Instead, their empirical manifestation follows from the manag-ers' accounts. Therefore, directing questions were shunned as much as possible, and in order to keep the interviews going, to redirect respondents to previous markers when they became repetitive, and to let the respondent elaborate on certain topics to ensure that his causal story was rooted in an empirical event (cf. Wuisman, 2005), different probes were used (Zeisel, 1981).

The interviews were transcribed and interpretively analyzed (Schwartz-Shea & Yanow, 2012) by applying qualitative coding techniques (Boeije, 2010) using AT-LASti software. More specifically, the transcripts were interpretively coded by discerning segments of text which were subsequently labeled and connected to one of the sensitizing concepts. This coding process is not linear but iterative. This means that through a process of memo writing and –reading (Schwartz-Shea & Yanow, 2012), drawing figures and matrices, discussions with colleagues, consulting literature and writing, codes were created and categorized and also renamed and re-categorized (Boeije, 2010). This coding process resulted in over 300 codes organized in 3 main categories: actions, motivations and situations (see Figure 1). The data could then be examined for patterns by e.g., examining what codes (i.e., issues and goals) regularly occurred and what codes regularly co-occurred (in an empirical event), thereby in effect linking the data to the in-terrelated sensitizing concepts of the analytical framework.

In addition, 50 documents including webpages were systematically collected via the website of the project (www.a2maastricht.nl) and also coded (Boeije, 2010) with ATLASti. This resulted in 217 codes organized in 5 categories (i.e., dates, actors, plans and studies, decisions, and substantive issues). This systematized document analysis resulted in a thick-case description that chronologically de-picts the development of the project in terms of involved actors, developed plans and studies, decision moments and substantive issues.

The document analysis is reported in the next section and focuses on the de-velopment of the project from the outset up to the construction phase. The empirical results of the interview analysis are reported in sections "Constituting Boundaries" and "Operations Of The Management Systems" and focus on the constitution and operations of the current management system.

The A2 Passageway Maastricht Infrastructure Project

The Dutch national highway A2 forms an east-west barrier in the city of Maastricht. Owing to increased traffic leading to problems such as con-gestion and rat-run traffic, from the 1980s to the late 1990s several at-tempts were undertaken to plan the reconstruction of the A2 Passageway and the northerly highway junction A2/A79. These failed mainly due to inadequate financial resources and a lack of political support.

The breakthrough came in 2001. From that moment on, governmental actors started cooperating and the scope of the project was broadened to include is-sues such as livability and safety. Rijkswaterstaat (RWS)—the executive arm of the current Dutch Ministry of Infrastructure and the Environment (I&M)—to-gether with the Province of Limburg and the Municipalities of Maastricht and

Meerssen, published the report 'Maastricht is Losing the Way' which led to re-newed financial negotiations between the national and regional governments resulting in an administrative agreement in January 2003. The national planning procedure 'Route Decision/Environmental Impact Assessment' started in June 2004 and the local planning procedures—the development of two zoning schemes—started in December 2004. Because of the interrelated nature of the project—the combined reconstruction of the highway A2, junction A2/A79, local roads, landscape and the urban area—it was decided that the planning procedures should be interwoven as much as possible. The national planning procedure led to the 'Funnel Decision' by the former Minister of Transport, Public Works and Water Management (V&W) that the Passageway was to be reconstructed as a tunnel. Based hereon, the four parties signed a cooperation agreement in 2006 and shortly thereafter the Municipal Council of Maastricht also positively decided on the tunnel variant.

Normally in the Netherlands these two decisions are first translated into a 'Design Route Decision' (OTB) on the national level and 'Design Zoning Plan' (OBP) on the municipal level, followed by public consultation procedures whereupon the responsible governments adapt their OTB or OBP into a 'Route Decision' (TB) or 'Zoning Plan' (BP) respectively, followed by the possibility for appeal at the Dutch Council of State. Thereafter, the procurement procedure would start resulting in a construction and adherent implementation plan. In this project this process was interwoven and inverted (cf. Van Valkenburg & Nagelkerke, 2006). In 2007 the public partnership published their 'Ambition Document' in which private consortia were asked to invent total solutions for the demarcated Maastricht area to solve interrelated problems regarding traffic congestion, accessibility, livability (e.g., rat-run traffic, air quality, noise pollution), safety, the barrier effect of the A2 and urban renewal. Thus, the procurement was a competition based on quality instead of budget; the budget was set by the public partnership (cf. Lenferink et al., 2011). Three consortia were selected after a first round and on the 25th of June 2009 the partnership announced that the plan 'The Green Ribbon' of consortium Avenue2 had won. The year 2010 was dedicated to integrating this plan into the OTB and OBP respectively, and subsequently the TB and BP. The national and local planning procedures truly synchronized on the 19th and 21st of November 2011 when they were treated jointly by the Council of State. The plans were declared irrevocably on the 30th of November 2011 and the implementation of 'The Green Ribbon' then formally commenced.

Interview Analysis

Constituting Boundaries: Two Management Systems

Based on the interview analysis of membership positions, interaction patterns and goals, two management systems can be distinguished: the public Project Agency A2 Maastricht and the private consortium Avenue2. For one thing, in terms of membership, public managers form part of the Project Agency A2 Maastricht and private managers belong to the consortium Avenue2. On the other hand, managers continuously interact both with people within their organization as with their 'counterpart' managers. This could point to the existence of a single management system. Moreover, the analysis clearly shows that four interrelated goals are regarded important by both public and private managers: time, budget, quality—in the literature known as the 'iron triangle' of project management or project success (Atkinson, 1999)—and integrality. It was expressed by all managers that time delays, budget overruns and low quality are highly undesirable (cf. Flyvbjerg *et al.*, 2003), and integrating infrastructure and urban development (and different related planning procedures) was considered imperative due to the integrated nature of the local urban layout, and expected by both public and private mangers to result in especially quality and time gains (cf. Lenferink *et al.*, 2011; Van Valkenburg & Nagelkerke, 2006).

However, the motivations expressed by public and private managers underlying these goals differ. This is mainly apparent from their understanding of quality and its interrelatedness with budget and time. For private managers time delays mainly threaten the business case of the project and ultimately the very existence of the construction companies, and quality is mainly understood in terms of scope: the number of things and their physical qualities that have to be built. For public managers time delays are mostly undesirable because of the extra hindrance (i.e., loss in quality) in terms of i.a. safety, traffic, noise, dust and vibrations that this causes; and budget overruns can cause, e.g., reputational damage. Hence, two management systems are distinguished.

The private system Avenue2 is a consortium of two large construction companies that are driven by the abovementioned main private motivation, and people from the two companies are mixed within divisions such as Design and Contract (see appendix). The public system seems less uniform in this respect. Managers originate from Rijkswaterstaat and the Municipality of Maastricht, which together form the formal principal of the project. Rijkswaterstaat managers primarily focus on the quality of products, contracts and procedures (see appendix). Their actions are aimed at risk minimization—which they term project management or system-oriented contract management—whereby they use tools such as risk databases and matrices. Municipal managers, on the other hand, are more concerned with informing and negotiating with societal actors in the environment and stress the importance of i.a. "healing the east and west" of the city and limiting hindrance for the local environment. However, respondents indicated that over time, as they jointly experienced the development of the project and are housed on the same floor of a single building, a public "project culture" emerged with a "single set of norms, values and principles". Motivations and goals became more integrated.

Operations Of The Management Systems

In line with the analytical framework depicted as Figure 1, it was analyzed what and how management actions were triggered by situations put forward by respondents and how these actions took shape. Table 1 provides a schematic overview of the most discussed situations in random order. The second column briefly describes the situation, the third column states what management system it concerns and the fourth column describes the management actions that were triggered by these situations.

Managing Complexity Through System Synchronization

As explained, supported by memo writing, literature consultation, etc. (Boeije, 2010; Schwartz-Shea & Yanow, 2012), the empirical data was examined for patterns to describe how management practices take shape in practice and to explain how these influence provisional outcomes in projects. That is the focus of this section. The first paragraph discusses, based on the document analysis and some additional interview data, the planning phase of the project. The subsequent paragraphs discuss the construction phase of the project.

In the planning phase local and national public planning procedures were interwoven with each other and also with the procurement procedures (cf. Van Valkenburg & Nagelkerke, 2006). Public managers claim to have reduced the duration of the project by a year owing to this planning strategy. Although interwoven, procedures did not merge: public and private activities were strictly divided. It is determined by law that public planning procedures are a governmental task and that public actors cannot intervene in private plan development by e.g., "cherry picking" between plans. However, in order to effectively interweave the different procedures, public managers and private developers kept interacting and working together on the basis of the mutually underscored importance of integrating infrastructure and urban development and a swift conclusion of the planning phase. As a result of this area-oriented planning approach (cf. Priemus, 2007), public and private managers claim synergy effects and time gains. In short, public and private management practices were fragmented, but at the same time the two management systems were (at least in part) integrated in terms of joint goals (most evidently integral area development) providing a co-operation basis, resulting in positive outcomes.

	Situation	Actor	Action
1	Landscape architect West 8 disagrees with the angle at which a bicycle tunnel should cross the underlying road network	Avenue2	Comply with wishes of West8 to prevent break
2	Subcontractors hedge risks and are not integrality-minded	Avenue2	Strict management of contracts, formalized cooperation
3	Substantive integrality and multidisciplinarity of the project lead to alignment problems between parts of the project	Avenue2	Interface management, people management, being reticent in changing planning and design
4	Land acquisition is not yet finished due to persistent landowners	Agency	Informing people, personal contact between managers or alderman and owners, sometimes expropriation
5	As people actually experience construction inconveniences they ask questions/complain and/or make requests	Both, but mainly Agency	Making deals with people, Communication Department, complaint hotline, consortium director interacts with societal environment, information evenings, informing people through website and weekly mailings, handing out cakes, informing political system
6	Vocational school Leeuwenborgh demanded that a temporary road was to be located further away from the school + financial compensation	Agency	The 'manager area' engaged in a transparent negotiation process and the alderman was put forward with extra money when his financial mandate proved insufficient
7	Water Board whishes changes in the project design	Both	Discussion and friction between Avenue2 and the Agency about who should pay for the changes

8	Disagreement between public stakeholders about tunnel safety requirements	Agency	Project director brings people informally together in service building of the Roermond tunnel (notorious to interviewees because of failures) and enforces mutual understanding and agreement
9	Cyclist Union complains about temporary diversions	Both	General director Avenue2 personally interacts with Union and plans are changed. Because of time pressure the municipal council was not informed about changes, leading to political commotion. The 'manager area' soothes emotions, prepares a meeting with the council which fizzles out.
10	Technical complexity of many interacting components and systems in tunnel	Avenue2	Engineering and design, subcontracting VTTI
11	Associations and property owners appeal at Council of State against TB/BP	Agency	Making deals with appellants (even during Council sitting), but also being distant/formal towards appellants during appeal period ("no negotiation, led the court judge")
12	National discussion about tunnel safety standards	Both	Uncertainty about standard led to discussion between principal and contractor to form a VTTI alliance to share risks. Alliance idea was abandoned when uncertainty became less. Instead, the contract will be amended and Avenue2 and the Agency will cooperate to implement design changes

Table 1 *Actions Triggered by Situations in the Construction Phase*

In the current construction phase of the project the public-private interaction was a central topic raised by the interviewees. On the one hand, public and private managers stressed and illustrated the close cooperation and trust between principal and contractor, the "openness and transparency", the "what is best for project" attitude, "going for win-win" and the Cooperation Principles that are part of the contract. Principal and contractor are established in the same building, physically detached from their respective public and private mother organizations, and they i.a. together run a single Communication Department (Table 1, row 5). This indicates integration. On the other hand, the relationship between principal and Avenue2 is rather fragmented in terms of management activities. Avenue2 is mainly involved in the controlled design and implementation of the project (Table 1, rows 2, 3 and 10) whereas the Project Agency is mainly concerned with managing relationships with the societal environment of the project (Table 1, rows 4, 5, 6, 8 and 11).

This fragmentation between principal and contractor exhibits parallels with what is known in the literature as project and process management. Edelenbos and Teisman (2008: 618-619) explain that "in a project approach, the assumption is that problems and solutions (…) are reasonably stable. This makes it possible to use project management techniques: a clear objective, a fixed schedule, clear preconditions, and an end product that is agreed on at the start. (…). Process management (…) is actually based on the assumption of dynamics and complexity in the interests and perspectives of many actors" and sees problem solving as a continuous interaction process with the environment. As shown in Table 1, the private action system is mainly (but not exclusively: rows 1, 5 and 9) involved in project management (rows 2, 3 and 10). In order to generate profits and conclude the project on time and within scope, the integral plan 'The Green Ribbon' is decomposed by Avenue2 in different areas of implementation, and risk control and minimization are central. The public action system mainly exhibits process management activities (rows 4, 5, 6, 8, and 11) since it is also and especially motivated by delivering quality in terms of creating a better socio-physical environment—during and after the construction of the project—in the city of Maastricht.

In short, the case shows fragmentation in terms of motivations and project and process management activities in both the planning and construction phases, and integration in terms of i.a. joint goals (time, budget, quality and integrality), cooperation and trust and openness. This dualism is depicted in Table 2.

Although process and project management are often described in the literature as mutually exclusive opposites (Edelenbos & Klijn, 2009), the case study shows that they are combined to produce e.g., time gains and synergy effects (cf. Edelenbos & Teisman, 2008). This does not imply a merging of public and private motivations, which may impede efficient and effective project realization (cf. Jacobs, 1994). However, because principal and contractor both benefit

	Dimension	Brief characterization
Fragmentation	Motivations (main)	Private: business case
		Public: quality for socio-physical system
	Management activities	Private: mainly project management (formal steering, contracts, risk minimization and control, planning)
		Public (Municipality): mainly process management (informal steering, interaction and communication, engagement with environment)
		Public (Rijkswaterstaat): mainly project management (system-oriented contract management, risk management)
Integration	Goals	Project realization within time and budget, quality, integrality
	Public-private relationship	Cooperation, trust, openness, win-win, best for project

Table 2 *Fragmentation and Integration*

from a project that is finished on time and within budget, and because principal and contractor both realize that both project and process management are imperative to realize this, activities can be fragmented without the project as a whole becoming disintegrated. However, as time and budget pressures in this phase of the project increase and the environment becomes very dynamic—stakeholders now start to 'feel' the inconveniences resulting from the construction—fragmentation seems to be stressed somewhat more. Managers of both systems discuss frictions in the cooperation regarding "additional work" (Table 1, row 7), e.g., disputes arise regarding who should bear the costs for changes in the construction plans.

Nevertheless, fragmentation is not a problem to be solved since it is necessary for managing complexity, i.e., for taking actions in complex systems and realizing goals effectively (cf. Jacobs, 1994). However, the success of this 'division of labor' depends on the ability to maintain a certain level of integration (Table 2), especially under conditions of increasing pressures from the systems' environment. In other words, fragmentation and integration need to be balanced.

This corresponds to what Teisman and Edelenbos term system synchronization "in which different subsystems largely hold their identity, but strive for coherence" (2011: 101). Teisman and Edelenbos (2011: 110) argue that system synchronization requires [1] self-organization and joint interests, [2] "satisfying the aims of self with the joint ambition", and the [3] "ability to reflect on and to redefine the boundaries of where actors are hold accountable". This article adds to this literature by showing that this involves [1] fragmentation of management activities and integration of goals, and [2] fragmentation of motivations and a good public-private relationship based on similar goals (Table 2). The interview analysis revealed that the principal-contractor and public-public cooperations are considered success factors in the project, and management activities are divided and shaped by private and public motivations respectively, but both have a strong interest in finishing the project on time and within budget. In the case of A2 Maastricht the third element identified by Teisman and Edelenbos (2011) is challenging. For instance, in the situation with the Water Board (Table 1, row 7), boundaries were not defined sufficiently clear in the planning phase. Respondents indicated ambiguity regarding who should have consulted the Water Board about their wishes related to the project. This resulted in disagreement between principal and contractor in the construction phase, were pressures on the project have increased, about who was responsible and thus should bear the costs of changing the plan design. The third element identified by Teisman and Edelenbos (2011) points to the importance of adaptation and flexibility (Edelenbos *et al.*, 2009)—reflecting on and redefining boundaries—for resolving conflict, i.e., maintaining system synchronization. This may require boundary spanning (Teisman & Edelenbos, 2011) and process management, especially in situations where managers are confronted with high pressures and complexity (e.g., Table 1, row 9) and the reflex is to fragmentize hence lessening synchronization (cf. Edelenbos & Teisman, 2008).

Discussion And Conclusion

The aim of this article was to show how management practices take shape in practice and to explain how these influence provisional outcomes in projects, thereby providing insights into the interaction of management with existing institutional contexts. For this purpose a complexity-informed research framework was developed and applied to the case of the A2 Passageway Maastricht infrastructure project. Although the framework allowed for the detailed grounded investigation of social systems and management behavior, it can be improved in further studies to be better able to analyze the dynamics of managing systems.

The analysis showed that both project and process management activities are performed in a combined fashion (cf. Edelenbos & Klijn, 2009; Edelenbos & Teisman, 2008). These activities interact with the institutional context by means of system synchronization (Teisman & Edelenbos, 2011) the act of 'balancing fragmentation of motivations' (cf. Jacobs, 1994) and management, with integration of goals and relationships. The case analysis indicates that this may positively affect the development of a project (e.g., quality and time gains) although under conditions of increasing contextual pressures and complexity, the focus seems to shift somewhat towards fragmentation.

E:CO Vol. 14 No. 4 2012 pp. 17-37

The analysis in this article has concentrated on the single case of the A2 Passage-way Maastricht project. Although this case study, supported by literature and insights from complexity science, hints at the importance of the balanced dualism identified here, it remains to be seen whether this patterns holds in a wider set of cases. Moreover, although the case study suggests that system synchronization decreases under conditions of increased complexity and system pressures, more case studies are needed to identify and conceptualize these conditions more precisely, to evaluate how these combine with certain management practices to produce certain outcomes (Verweij & Gerrits, forthcoming), and to see how such patterns evolve. The results of such comparative analyses can subsequently be fed back into individual cases such as the A2 Maastricht project to increase our understanding of how complexity manifests itself empirically and what management practices can be deployed under specific contextual conditions to cope with or even harness it.

Acknowledgements

I am indebted to Geert R. Teisman for his constructive feedback in the development of this article. I am also grateful to the reviewers for their comments which helped me to make improvements to the article. This work was supported by Next Generation Infrastructures (grant 03.24EUR). A previous version of this article was presented at the 7th International Conference in Interpretive Policy Analysis in Tilburg, the Netherlands.

Appendix: Interview Data

Nr.	Date	Respondent	Organization	Included	Notes
1	2011-09-12	Manager A2 School	Municipality of Maastricht	No	Pilot interview, gain access to project, interview strategy did not fit research approach of paper, no manager of project
2	2011-09-13	General director	Avenue2	No	Pilot interview, interview strategy did not fit research approach
3	2011-09-23	Manager planning	Project Agency (RWS)	No	Idem
4	2011-09-23	Manager area	Project Agency (Maastricht)	No	Idem
5	2011-09-23	Manager integral area development	Project Agency (RWS)	No	Idem
6	2011-11-14	Manager contract	Project Agency (RWS)	Yes	Location: Utrecht Duration: 86 minutes
7	2011-11-18	Manager design	Avenue2	Yes	Location: Nieuwegein Duration: 101 minutes
8	2011-11-18	Manager VTTI	Avenue2	Yes	Location: Nieuwegein Duration: 144 minutes
9	2011-11-21	Manager contract	Avenue2	Yes	Location: Maastricht Duration: 69 minutes

10	2011-11-21	Manager integral area development	Project Agency (RWS)	Yes	Location: Maastricht Duration: 75 minutes
11	2011-11-22	Manager business operations	Project Agency (Hired)	Yes	Location: Maastricht Duration: 85 minutes
12	2011-11-23	General director	Avenue2	Yes	Location: Maastricht Duration: 43 minutes
13	2011-11-23	Manager area	Project Agency (Maastricht)	Yes	Location: Maastricht Duration: 60 minutes
14	2011-11-23	Project director	Project Agency (Maastricht)	Yes	Location: Maastricht Duration: 66 minutes
15	2011-11-28	Manager GWW	Avenue2	Yes	Location: Maastricht Duration: 72 minutes
16	2011-11-29	Chief work planning	Avenue2	No	No manager of the system central to this paper, too strict focus on internal team
17	2011-11-29	Manager civil	Avenue2	Yes	Location: Maastricht Duration: 93 minutes
18	2011-12-05	Manager planning	Project Agency (RWS)	Yes	Location: Maastricht Duration: 112 minutes
19	2011-12-06	Manager tender	Avenue2	Yes	Location: Utrecht Duration: 77 minutes

References

Atkinson, R. (1999). "Project management: Cost, time and quality, two best guesses and a phenomenon: It's time to accept other success criteria," *International Journal of Project Management*, ISSN 0263-7863, 17(6): 337-342.

Boeije, H. (2010). *Analysis in Qualitative Research*, ISBN 9781847870070.

Boulton, J. (2010). "Complexity theory and implications for policy development," *Emergence: Complexity & Organization*, ISSN 1532-7000, 12(2): 31-40.

Bowen, G.A. (2006). "Grounded theory and sensitizing concepts," *International Journal of Qualitative Methods*, ISSN 1609-4069, 5(3): 12-23.

Buijs, M.J., Eshuis, J. and Byrne, D.S. (2009). "Approaches to researching complexity in public management," in G.R. Teisman, M.W. van Buuren and L.M. Gerrits (eds.), *Managing Complex Governance Systems: Dynamics, Self-Organization and Coevolution in Public Investments*, ISBN 9780415459730, pp. 37-55.

Byrne, D.S. (1998). *Complexity Theory and the Social Sciences: An Introduction*, ISBN 9780415162968.

Byrne, D.S. (2005). "Complexity, configurations and cases," *Theory, Culture & Society*, ISSN 1460-3616, 22(5): 95-111.

Byrne, D.S. (2011). *Applying Social Science: The Role of Social Research in Politics, Policy and Practice*, ISBN 9781847424501.

Checkland, P. (1981). *Systems Thinking, Systems Practice*, ISBN 9780471986065.

Cilliers, P. (2001). "Boundaries, hierarchies and networks in complex systems," *International Journal of Innovation Management*, ISSN 1363-9196, 5(2): 135-147.

Dennard, L.F., Richardson, K.A. and Morçöl, G. (eds.). (2008). *Complexity and Policy Analysis: Tools and Concepts for Designing Robust Policies in a Complex World*, ISBN 9780981703220.

Edelenbos, J. and Klijn, E.H. (2009). "Project versus process management in public-private partnership: Relation between management style and outcomes," *International Public Management Journal*, ISSN 1559-3169, 12(3): 310-331.

Edelenbos, J. and Teisman, G.R. (2008). "Public-private partnerships on the edge of project and process management: Insights from Dutch practice: The Sijtwende spatial development project," *Environment and Planning C: Government and Policy*, ISSN 0263-774X, 26(3): 614-626.

Edelenbos, J., Klijn, E.H. and Kort, M.B. (2009). "Managing complex process systems: Surviving at the edge of chaos," in G.R. Teisman, M.W. van Buuren and L.M. Gerrits (eds.), *Managing Complex Governance Systems: Dynamics, Self-Organization and Coevolution in Public Investments*, ISBN 9780415459730, pp. 172-192.

Elder-Vass, D. (2005). "Emergence and the realist account of cause," *Journal of Critical Realism*, ISSN 1572-5138, 4(2): 315-338.

Fischer, F. (1998). "Beyond empiricism: Policy inquiry in postpositivist perspective," *Policy Studies Journal*, ISSN 1541-0072, 26(1): 129-146.

Flyvbjerg, B., Bruzelius, N. and Rothengatter, W. (2003). *Megaprojects and Risk: An Anatomy of Ambition*, ISBN 9780521009461.

Gerrits, L.M. (2008). *The Gentle Art of Coevolution: A Complexity Theory Perspective on Decision Making Over Estuaries in Germany, Belgium and the Netherlands*, Rotterdam: Erasmus University Rotterdam.

Gerrits, L.M. and Verweij, S. (forthcoming). "Critical realism as a meta-framework for understanding the relationships between complexity and qualitative comparative analysis," *Journal of Critical Realism*, ISSN 1572-5138,

Jacobs, J. (1994). *Systems of Survival: A Dialogue on the Moral Foundations of Commerce and Politics*, ISBN 9780679748168.

Lenferink, S., Arts, J. and Tillema, T. (2011). "Ongoing public-private interaction in infrastructure planning: An evaluation of Dutch competitive dialogue projects," in K.V. Thai (ed.), *Towards New Horizons in Public Procurement*, Boca Raton: PrAcademics Press, pp. 236-272.

Meadows, D.H. (2008). *Thinking in Systems: A Primer*, ISBN 9781603580557.

Meek, J.W. (2010). "Complexity theory for public administration and policy," *Emergence: Complexity & Organization*, ISSN 1532-7000, 12(1): 1-4.

Mingers, J. (2003). "Observing organizations: An evaluation of Luhmann's organization theory," in T. Bakken and T. Hernes (eds.), *Autopoietic Organization Theory: Drawing on Niklas Luhmann's Social System Perspective*, ISBN 9788763001038. pp. 103-122.

Moeller, H. (2006). *Luhmann Explained: From Souls to Systems*, ISBN 9780812695984.

OECD Global Science Forum. (2009). *Applications for Complexity Science for Public Policy: New Tools for Finding Unanticipated Consequences and Unrealized Opportunities*, Paris: OECD.

Priemus, H. (2007). "System innovation in spatial development: Current Dutch approaches," *European Planning Studies*, ISSN 1469-5944, 15(8): 991-1006.

Reed, M. and Harvey, D.L. (1992). "The new science and the old: Complexity and realism in the social sciences," *Journal for the Theory of Social Behavior*, ISSN 1468-5914, 22(4): 353-380.

Rhodes, M.L., Murphy, J., Muir, J. and Murray, J.A. (2011). *Public Management and Complexity Theory: Richer Decision-Making in Public Services*, ISBN 9780415457538.

Schwartz-Shea, P. and Yanow, D. (2012). *Interpretive Research Design: Concepts and Process*, ISBN 9780415878081.

Seidl, D. (2005). "The basic concepts of Luhmann's theory of social systems," in D. Seidl and K.H. Becker (eds.), *Niklas Luhmann and Organization Studies*, ISBN 9788763001625, pp. 21-53.

Stacey, R.D. (1992). *Managing the Unknowable: Strategic Boundaries Between Order and Chaos in Organizations*, ISBN 9781555424633.

Teisman, G.R. and Edelenbos, J. (2011). "Towards a perspective of system synchronization in water governance: A synthesis of empirical lessons and complexity theories," *International Review of Administrative Sciences*, ISSN 1461-7226, 77(1): 101-118.

Teisman, G.R. and Klijn, E.H. (2008). "Complexity theory and public management: An introduction," *Public Management Review*, ISSN 1471-9045, 10(3): 287-297.

Teisman, G.R., Gerrits, L.M. and Van Buuren, M.W. (2009a). "An introduction to understanding and managing complex process systems," in G.R. Teisman, M.W. van Buuren and L.M. Gerrits (eds.), *Managing Complex Governance Systems: Dynamics, Self-Organization and Coevolution in Public Investments*, ISBN 9780415459730, pp. 1-16.

Teisman, G.R., Van Buuren, M.W. and Gerrits, L.M. (eds.) (2009b). *Managing Complex Governance Systems: Dynamics, Self-Organization and Coevolution in Public Investments*, ISBN 9780415459730.

Uprichard, E. and Byrne, D.S. (2006). "Representing complex places: A narrative approach," *Environment and Planning A*, ISSN 0308-518X, 38(4): 665-676.

Van Valkenburg, M. and Nagelkerke, M.C.J. (2006). "Interweaving planning procedures for environmental impact assessment for high level infrastructure with public procurement procedures," *Journal of Public Procurement*, ISSN 1535-0118, 6(3): 250-273.

Verweij, S. and Gerrits, L.M. (forthcoming). "Understanding and researching complexity with qualitative comparative analysis: Evaluating transportation infrastructure projects," *Evaluation*, ISSN 1461-7153,

Wagenaar, H. and Cook, S.D.N. (2003). "Understanding policy practices: Action, dialectic and deliberation in policy analysis," in M.A. Hajer and H. Wagenaar (eds.), *Deliberative Policy Analysis: Understanding Governance in the Network Society*, ISBN 9780521530705, pp. 139-171.

Waldrop, M.M. (1992). *Complexity: The Emerging Science at the Edge of Order and Chaos*, ISBN 9780671872342.

Weiss, R.K. (1994). *Learning from Strangers: The Art and Method of Qualitative Interview Studies*, ISBN 9780684823126.

Wuisman, J.J.J.M. (2005). "The logic of scientific discovery in critical realist social scientific research," *Journal of Critical Realism*, ISSN 1572-5138, 4(2): 366-394.

Zeisel, J. (1981). *Inquiry by Design: Tools for Environment-Behavior Research*, ISBN 9780521319713.

Stefan Verweij M.Sc. is a Ph.D. Student at the Department of Public Administration at Erasmus University Rotterdam in the Netherlands. Within that department, he is involved in the research group Governance of Complex. His current research focusses on the development of complexity-informed methodologies for evaluating infrastructure and spatial planning projects. More information about his research, publications and contact details can be found at his website: www.stefanverweij.eu.

Applied

A Systems Perspective On U.S. Foreign Policy In The Middle East: A Propositional Analysis

Peter Daniels
Swarthmore College, USA

This paper explores the contribution that complexity science might play in the development of U.S. foreign policy in regard to the Middle East. Three cases grounded in history, language, culture, and other social phenomena illustrate the extent to which United States policy actors are challenged to grasp the nature of foreign nations as complex systems. The different issues in each case—a linguistic case in Egypt, a religious/political case in Lebanon, and religious/political/economic case in Saudi Arabia—display relationships characteristic of complex systems that substantially affected each situation. Five propositions are then offered as potential means to counteract this tendency to organize policy responses on a simplified cognitive frame, thereby giving complexity science a larger role in the development of U.S. foreign policy.

Introduction

The United States, like other western nations, has not been hugely successful in enacting a consistent and coherent foreign policy in the Middle East. The tensions and conflicts between European nations and their Middle Eastern counterparts date back at least to the year 1095 AD, when Pope Urban II called for the military expedition that became known as the First Crusade (Bongars, 1611 as in Thatcher and McNeal, 1905), which was followed by six more Crusades over the next two centuries. The modern policy era seems to be faring little better, given the recent wars in Iran, Kuwait, Israel/Palestine, Iraq, and Afghanistan. The persistent lack of effective engagement has led some analysts to argue that discord between civilizations is redefining the world order (Huntington 1997).

The thesis of this article is that the foreign policy of the United States has not been grounded in a sufficiently sophisticated systems perspective of the Middle East, and, by extension, that improving that systems perspective offers the possibility of at least improving our ability to achieve our foreign policy goals. The Middle Eastern nations represent diverse cultures that are unique assemblages of language, religion, history, politics, and economic forces. U.S. foreign policy in the region has arguably failed to recognize the systemic diversity between the nations (although Iraq and Iran share a border, they could scarcely be more different), but it has also failed to recognize how the social systems in the Middle East are dramatically different than those of the Anglo-European model, implying that a Euro-centric paradigm is likely to be ill-suited to the region. The point of the departure for this thesis is the assumption that U.S. foreign policy actors engage in a fundamental attribution error (Thompson *et al.,* 2004) by failing to recognize how truly different the experiences of those in Muslim countries in the Middle East are from their own lived experiences and worldviews. As such, there is a predisposition to reason by analogy and attempt to apply Western thought processes and values to the region. Former Secretary of State Condoleezza Rice said in a 2005 speech in Cairo:

> *America was founded by individuals who knew that all human beings—and the governments they create—are inherently imperfect. After all, the United States was born half free and half slave. And it was only in my lifetime that my government guaranteed the right to vote for all of its people. Nevertheless, the principles enshrined in our constitution enable citizens of conviction to move us ever closer to the ideal of democracy. Here in the Middle East, the long, hopeful process of democratic change is now beginning to unfold. Millions of people are demanding freedom for themselves and democracy for their countries.* (Rice, 2005)

Implicit in the logic of this passage is the assumption that "the principles enshrined in our constitution" are relevant in the Middle East, and that whatever "freedom" and "democracy" that might emerge in the region would somehow be similar to those constructs in the US.

This paper progresses through two phases. First, a set of cases illustrates how different a Middle Eastern policy context is from an Anglo-European context. These cases show that a simplistic understanding of language, policy structures, and religion can lead to misunderstandings and unintended consequences. Second, a set of propositions extends the systems argument into the likely impacts on policy structures and outcomes one might reasonably expect from a more systems-grounded approach. The three cases chosen for this paper illustrate the complexity of the situations in the Middle East into which the United States inserts itself. The most significant foreign policy decisions of the United States regarding the Middle East have dealt with circumstances that manifest this complexity. These complex settings are in fact systems constituted of interactions between language, culture, politics, social values, and an often-vast number of diverse stakeholders. In the past, instead of being appropriately grounded in cultural complexity, U.S. policy responses to different situations have been too focused on a single issue or facet of the problem at hand (Frej & Ramalingam, 2011). Of course, they have also always been shaped by factors other than the apparent situation in the Middle East, such as domestic politics, funding difficulties, etc.

Cases

The following set of illustrative examples was chosen with the intention of displaying historical, geographical, and thematic cross-sections of the patterns characteristic of the region. Although each situation deals with different phenomena, societies, regions, and actors, their commonalities display the complex nature of the systems with which the United States interacts. There are certainly more complex policy dilemmas in the Middle East (women's rights, religious tolerance, etc.) but a full elaboration of the policy implications of their systems complexity is beyond the scope of this paper.

Case 1: Linguistic Nuance in Egypt

The first case illustrating the complex nature of societal systems in the Middle East is found in Egypt. As both a major player in the Middle East and one of the hotspots of the 2011 Arab Spring protests, Egypt has become especially and increasingly important to both U.S. media and U.S. policymakers since the events of 25 January 2011. However, as news corporations updated the rest of the world, they tended to mention a few themes repetitively: terrorism, Islamism, the Muslim Brotherhood, the status of women, etc. Often lost in the analysts' sound-bite statements attempting to explain each day's developments were the subtle, more nuanced cultural phenomena at play.

This was brought into stark clarity by the U.S. media's struggle to understand the popular Egyptian views of the protestors in Tahrir Square following 25 January. To be clear, the situation was very dynamic and their confusion was clearly not unfounded. Crowds of protesters were ballooning in Tahrir and, simultaneously, the low-income citizens across the city who had the most to gain from the revolution were ironically accusing the protestors of creating instability and destroying the economy. While much of this misapprehension was eventually pinned on government propaganda, the language used by Egyptians on the street included a few key linguistic phenomena that were lost in translation to the untrained ear. The clearest example of this was the representation of violence by protestors and their opposition (Ghannam, 2012).

Two Arabic words that are often associated with violence and those who engage in it are *gad'ana* and *baltaga*. Both of these words have sometimes been translated simply as "violent" or "violence." However, they have starkly different connotations. Since there are no direct equivalents in English, explaining their meaning is a difficult task at best. Sawsan El-Messiri called gad'ana a combination of "nobility, audacity, responsibility, generosity, vigor, and manliness" (1978: 49). But Farha Ghannam also noted earlier this year that "part of gad'ana is the ability to use measured violence in proper contexts, including self-defense and to help or protect vulnerable individuals and family members (especially females)" (2012: 33-34). Thus, gad'ana must be distinguished from nonviolence.

E:CO Vol. 14 No. 4 2012 pp. 38-50

This further reflects the distinction between Jesus's command to "turn the other cheek" and the Qur'an's mandate to forcefully defend one's people against oppression.

In contrast, baltaga is violence that is used inappropriately, i.e. to bully, oppress, or impose one's will on another. This phrase is most often used in reference to *baltagiyya*, or thugs, a construal that dates back to gangs of knife-wielding soldiers protecting the Ottoman sultan in the 12th Century. The rhetorical frequency of baltagiyya was rare in many neighborhoods in Cairo, until, in the days following the protests, a group of men on camel- and horseback claiming to be supporters of the Mubarak regime rode into the crowds of protesters with swords, axes, and guns, becoming an almost too perfect representation of baltaga.

After this event, public perceptions of the protesters and of their opposition began to shift. The protestors were seen less as powerless troublemakers, as the government was trying to suggest, and more as an active group of gid'aan (those who embody gad'ana). The protestors' nonviolent methods balanced with their bravery in the face of oppression painted an ideal picture of a new type of Egyptian hero. Mubarak, on the other hand, was attached to a culturally embedded image of thuggery, corruption, and unnecessary violence. But in news accounts, both gad'ana and baltaga could both be translated as violence. Because English does not have a corresponding linguistic distinction, failing to distinguish between them would be about as accurate as comparing Batman to the Oklahoma City bomber.

Case 2: Lebanese Sectarianism

The next case moves northeast to Lebanon. Although the contemporary state of Lebanon is a relatively recent innovation that is bounded by Western-drawn borders, it was a region defined by sectarianism for years earlier, and by religious diversity for millennia before that. Furthermore, it is undeniably important to the United States in its current relationships to Israel/Palestine and Hezbollah, as it was in 1958 when 5,000 U.S. Marines landed on the Lebanese coast to defend a Christian regime.

Since the United States is clearly interested in Lebanon and the role of religion therein, it is therefore crucial for U.S. policymakers to understand sectarianism and its relationship to politics. Currently, the Parliament of Lebanon is a unicameral, formally sectarian body, with the 128 seats apportioned between the Christian and the Muslim communities in Lebanon (with some diversity in terms of specific confession—Shi'a, Sunni, Druze, Maronite, Orthodox, etc.).

However, many Western thinkers would like to conceptualize sectarianism in Lebanon as a primordial force that has always defined cultural interactions in Lebanon and the Middle East as a whole. This is not historically accurate; to the contrary, Lebanon was known as a refuge for persecuted religious groups for centuries before it became a Western-style nation-state. As much as Western thinkers would like to think of the recent religious tensions dating back millennia, Lebanese sectarianism as it is known today is, like the modern state of Lebanon itself, a relatively recent innovation. Makdisi put it best:

> It [sectarianism] is, rather, an intermingling of both precolonial (before the age of Ottoman reform) and postcolonial (during and after the age of reform) understandings, metaphors, and realities that has to be dissected at at least two overlapping and mutually reinforcing levels, of the elite and nonelite. In other words, sectarianism is a modernist knowledge in the sense that it was produced in the context of European hegemony and Ottoman reforms and because its articulators at a colonial (European), imperial (Ottoman), and local (Lebanese) level regarded themselves as moderns who used the historical past to justify present claims and future development. (2000: 7)

This analysis is evidence that the political and cultural landscapes of regions of the Middle East cannot be isolated from their history, religion, and government, nor their manifold interactions with a powerful West.

In 1958, there was a substantial Muslim uprising against the Lebanese Christian president. Under the auspices of the Eisenhower doctrine, the United States came to the rescue of the Christian administration, and landed 5,000 U.S. Marines on the coast around Beirut. The uprising was put down, but at the cost of an already weak base of sympathy for the U.S. from a vastly Muslim Arab world. This bold action, coupled with the United States's unwavering support of Israel, put it fundamentally at odds with a majority of the Middle East. But specifically, these historical moments are just a glimpse into the complex political, social, and religious landscapes of Lebanon that interact and engage with each other to form a regional culture (Rogan, 2009; Cleveland & Bunton, 2009).

Case 3: The Expansion of Wahhabism

A form of Salafism, Wahhabism is an Islamic sect founded in Saudi Arabia by Muhammad ibn 'Abd al-Wahhab in the early 18th Century. The Wahhabis can be regarded as the puritan evangelicals of Islam. Their most distinguishing and fundamental ideological feature is their unrelenting and incredibly strict focus on the doctrine of *tawhid,* loosely translated as "the oneness of God." This, complemented by a very literal interpretation of the Qur'an and a selective reading of *hadith* (records of the sayings of the Prophet Muhammad's *sunnah,* or sayings and actions) has resulted in a widespread Wahhabi rejection of anything that can be represented, or misrepresented, as *shirk,* or idolatry. They have become the staunch opponents of Sufism, a more mystical, esoteric, and popular expression of Islam found worldwide.

In 1740 ibn 'Abd al-Wahhab formed an alliance with Muhammad ibn Saud, then the ruler of a portion of what is now Saudi Arabia. Over the next century and a half, the Saud family would mount a conquest of most of the Arabian Peninsula. The Saudi family remains the royal family of Saudi Arabia today, and Wahhabism is still strongly espoused by the government. This has created a unique situation regarding the fundamental integration of church and state. Many in the United States are trained from an early age to preach "separation of church and state," so having policymakers recognize even the direct impact of religion on the state can prove difficult. (The Rice quote earlier in this paper illustrates this— her statement presupposes that a government is made solely by its people. To a conservative Wahhabi, this is difficult to conceive of). This problem is even more pronounced when religion and state are almost indistinguishable.

This apparent cultural divide became monumentally important with Saudi Arabia's 1933 oil concession to the U.S. Standard Oil of California (later ARAMCO) in preference to an Iraqi competitor. Following that agreement, the United States became not only a major player in the oil game with Saudi Arabia, but also became one of the Saudi Arabian government's larger funders. Of course, this oil-centric relationship has continued to today.

Now, Saudi Aramco, owned by the Saudi Arabian government, has been valued as the most profitable single business venture in the world. This has translated directly to strong financial support of Wahhabi evangelical efforts. With some estimates ranging as high as $100 billion in the last few years, the relatively small Wahhabi sect is known to control roughly 95% of Islam's total funds worldwide. Around 1,500 mosques worldwide are estimated to have been built using Saudi/Wahhabi funds in the last 50 years.

Ironically, the United States has put itself in a kind of double bind. By relying to an extent on foreign oil, specifically Saudi Arabian oil, it has funded the now expansive Saudi treasury. This, in turn, funds the efforts and expansion of Wahhabism, a group many Americans would call ideologically antithetical to traditional American values in multiple ways. American critiques of Wahhabism have often keyed on their historical reliance on violence and contemporary intolerance of much religious diversity. But despite this apparent cultural divide, the American oil economy continues to fund Wahhabi efforts (Rogan, 2009; Cleveland & Bunton, 2009).

Each of these three cases illustrates a similar dynamic. Societies, states, and organizations in the Middle East are complex systems, where minute or subtle cultural exchanges—such as the use of certain words or the development of a particular strain of Islam—can create sweeping structural changes. Due to this progression, it is particularly easy for external actors such as the United States to ignore not only the smaller cultural details, but the larger societal forces as well, sometimes at their own expense.

Propositions

If the operating premise of this article is correct—that U.S. policy initiatives and diplomatic interactions with Middle Eastern countries would be more successful if they were more informed by systems thinking concepts—the implication is that there should be a number of specific ways in which systems thinking might contribute to that success. Several key links between systems thinking and U.S. foreign policy in the Middle East are presented below as a series of propositions. It is important to note that these are framed in an incremental manner (increasing or decreasing progress.) It would be overly simplistic to view our current policy efforts as wholly unsuccessful or systems thinking as a panacea that will overcome the challenges created by the cultural differences.

Proposition 1: Policy structures that conceptualize culture as an emergent property are likely to experience more progress.

Probably the first day in "how to do foreign policy" class, the instructor pronounces that policy interventions must be culturally nuanced in order to be successful. Taking that as a given, the more interesting question becomes how best to conceptualize "culture." Two systems concepts—emergence and autopoiesis—offer interesting frames within which to locate a discussion of culture. Corning extends Holland's chess analogy (1998) in a useful way, when applying emergence to the notion of culture:

> *The game of chess illustrates precisely why any laws or rules of emergence and evolution are insufficient. Even in a chess game, you cannot use the rules to predict "history"—i.e., the course of any given game. Indeed, you cannot even reliably predict the next move in a chess game. Why? Because the "system" involves more than the rules of the game. It also includes the players and their unfolding, moment-by-moment decisions among a very large number of available options at each choice point. The game of chess is inescapably historical, even though it is also constrained and shaped by a set of rules, not to mention the laws of physics. Moreover, and this is a key point, the game of chess is also shaped by teleonomic, cybernetic, feedback-driven influences. It is not simply a self-ordered process; it involves an organized, "purposeful" activity.* (Corning, 2002: 14)

The expression of culture in the Middle East is a chess game of monumental proportions. The massive number of individuals, organizations, and institutions easily gives rise to innumerable cultural phenomena, which, in turn, create a regional culture—an "organized, 'purposeful' activity." Holland (1998) also introduces what might be referred to as the "enigma test": when phenomena appear enigmatic, it may be that the observer is functioning at the wrong scale, and failing to observe key emergent properties. Middle Eastern cultures certainly appear enigmatic to Western policy actors.

Autopoiesis, as articulated by Varela and Maturana (Maturana & Varela, 1980; Varela, Maturana, & Uribe 1974) is similarly applicable to understanding the factors that shape foreign policy success. Although the concept was initially developed to explain biological rather than social processes, viewing culture as an autopoietic system is rich with insights. Autopoiesis describes the property of being both the producer and the product; of self-reproduction and resilience; of autonomy and adaptation. To the extent that nation-states are autopoietic, they are self-correcting organizations that tend to defy external attempts to change them (e.g., resistance to environmental perturbations). The strong networks of linguistic, religious, and political structures in Middle Eastern countries make it seem likely that autopoiesis is a useful lens through which to view the region. At the very least, autopoiesis would lead one to adopt a longer time horizon within which to expect authentic systemic change.

Proposition 2: Highly compartmentalized policy structures are likely to experience less progress.

To the extent that US policy interactions are distributed across separate agencies and administrations, they are less likely to function as a coherent, mutually assisting system. Having education assistance programs administered through one agency, economic assistance through another, and security technical assistance through yet a third might be consistent with how our administrative structures are funded, but are substantially incompatible with the policy demands of effective interaction with Middle Eastern cultures. The Cartesian reductionist tendency to divide the systemic issue into separate problems must be rejected in favor of a strongly systems-based approach. Two U.S. scholars agree that:

> *Such conventional and reductionist ways of analyzing and dealing with social and economic problems help break down difficult undertakings, but they don't help deal with the complexity, uncertainty or ambiguity that characterizes the emergent behavior these complex adaptive problems display. This way of thinking has led many in foreign policy (and indeed domestic policy, although that is a separate set of issues) to act as though they can predict and precisely manage the behavior and outcomes of these systems.* (Frej & Ramalingam, 2011)

Proposition 3: Policy structures that can engage in Argyris-style "double-loop learning" are likely to experience more progress.

Chris Argyris was an early thinker about the importance of learning organizations (Argyris, 1962, 1964, 1965; Argyris & Schön, 1974; Argyris, 1982). Even if individuals within U.S. foreign policy agencies might have high levels of individual competence interacting with Middle Eastern cultures, that does not ensure that the organizations themselves can reflect a similar level of sensitivity and competence. Two of the concepts from the double-loop learning literature that are particularly relevant in this context are the openness to learning from errors and the willingness to reflect on the underlying assumptions—those held at both at the individual and organizational level—that may have contributed to those errors. Being open to reflecting on one's underlying assumptions is critical in intercultural systems because culture can be thought of as a taken-for-granted shared reality. That "taken-for-grantedness" makes culturally based assumptions notably invisible, durable, and socially reinforced.

Proposition 4: Policy structures with problematic hierarchical boundary conditions are likely to experience less progress.

Almost without exception U.S. policy structures are characterized by professional/career strata with an overlay of political appointees. Each presidential administration installs political appointees that are loyal to it self, and the appointees then function as agents of the administration's policy priorities and political ideology. So an embassy might be staffed by career employees, often with considerable in-country experience, while the ambassador is more likely to be a political appointee with much less experience or understanding of the nuances of the local context. This hierarchical structure—that day-to-day functionaries in an organization have more experience than higher status colleagues—is not unique to foreign policy. Large naval vessels are run by non-commissioned officers who might spend their entire career on the same ship while the officer structure is populated with upwardly mobile employees who may serve only one tour on any particular vessel. Surgical teams consist of nurses who do not have status comparable to the surgeons they assist, regardless of their comparative experience.

But how well these various hierarchically differentiated human activity systems function depends on the boundary conditions that emerge between the functional/low status members and the administrative/high status members. Organizations that minimize the communication and authority disconnects caused by the power distance between the two groups are more likely to be successful (Hofstede, 2001). Organizations that can minimize the emergence of destructive in-group/out-group social psychological dynamics between the strata are more likely to be successful (Fiske & Taylor, 2007). Again, taking a human activity systems perspective is a means to achieving this goal.

Proposition 5: Policy structures that conceptualize Middle Eastern states as evolutionary systems will be more adaptable to dynamic situations.

Cultural systems are the epitome of dynamic systems—the sheer number of social and ideological exchanges occurring daily makes this unquestionable. As these exchanges occur, not only does the system reinforce itself, but it also changes, often drastically, and on fundamental levels. A systems-based approach is a powerful way to account for this in policy structure development: "the ideas of the 'complexity sciences' could, in different contexts, greatly contribute to significant breakthroughs in how foreign policy efforts are analyzed, proposed and implemented and how foreign policy decisions are actually made" (Frej & Ramalingam, 2011). The conceptualization of social systems as evolutionary further allows more nuanced approaches to guide and effectively impact external interventions (Bánáthy, 1998). Systems thinkers such as Bánáthy offer extensive insight into how large institutions can utilize these concepts to effect real change on a societal level.

Extending The Propositions Into Practice

Assuming these propositions are at least in some part accurate, the logical next step is to ask, "So what? What kinds of methods or strategies could we employ to utilize these propositions and expedite their effects?" While the results of accepting the above propositions will include the intangible benefits of a different mode of thought, there are also many concrete ways to access their advantages. Proposition 1 (culture is emergent) not only leads one to have a more realistic and complete sense of a particular cultural phenomenon, but also could lead policymakers to recognize the need for broader cultural diplomacy programs with a much longer time frame than is often expected. The emergent nature of culture makes it resistant to narrower, targeted invasions and often frustrates early results of foreign policy. The recognition of culture as emergent would explain this delay and create a more realistic mindset regarding systemic cultural shifts. A very tangible operationalization of this proposition would be to think of developing "cultural fluency," rather than "language fluency." Proposition 2 (compartmentalism is limiting) will ideally lead policymaking teams and committees to eschew specific, compartmentalized subcommittees and turn instead to a kind of structural and functional collaboration on a system-wide level, accepting connections between disciplines as expected and useful. There is an argument in policy sciences that to really understand what organizations value, simply follow the money (Wildavsky, 1964). Adopting this proposition could lead to appropriating project money into pools that agencies can only access if they submit complexity-informed proposals that are collaborative across traditional agency missions.

Proposition 3 (learning leads to progress) lays at the core of making durable, far-reaching improvements in the practice of foreign policy. Organizations could, through engagement in double-loop learning processes such as self-reflection and learning from errors, encourage a greater ability on an organizational level to root out false underlying assumptions regarding other cultural systems. The foreign policy community could more quickly embrace learning as a fundamental organizational value by looking outside their own field for models and mentors. The most probable source of ideas and models is the field of contemporary environmental conflict management. Environmental conflict mediators and policy-makers have long been treating the environment as a system with all the properties mentioned elsewhere in this paper. Furthermore, myriad techniques, frameworks, and approaches have been constructed to develop policy accordingly (Daniels & Walker, 2001; Röling & Wagemakers, 1998; and Keen *et al.*, 2005). Organizations such as the Santa Fe Institute have been generating academic-to-practitioner dialogues that are a seedbed for organizational innovation and would be a useful think-tank resource for a policy-making team.

Proposition 4 (hierarchical boundaries can be obstructive) conflicts with multiple structural realities of the U.S. foreign policy world. It is unrealistic to challenge the political reality of ambassadors and other high officials being political appointees. However, a movement to lengthen the terms of those officials in a single region or country would substantially increase their ability to improve as experts in foreign policy and the area in question. Furthermore, such a change would encourage politicians to appoint as ambassadors candidates that will be most likely to succeed in the long term. Proposition 5 (Middle Eastern states are evolutionary systems) would likely lead to the rejection of more stagnant representations of Middle Eastern states. Just slightly outdated reports can easily become a touchstone for policymakers even when the report may be partially or entirely obsolete. A more dynamic, personal engagement with the issues by the appropriate officials would lead to the conceptualization of Middle Eastern states as both independent and extremely interdependent with each other and numerous other influential actors, the U.S. included. The easiest way to improve this would be to encourage or mandate increased direct interactions between the policymakers and the situation and culture at hand. The adoption of Proposition 4 would complement this, giving high-level officials the time and ability to view each situation as dynamic, instead of getting a shorter, snapshot view of the region.

Conclusions

These propositions are not meant to indict the entire system of U.S. foreign policymaking in the Middle East. Certainly Checkland's (1981) recommendation that the appropriate goal when intervening in complex human activity systems is *situation improvement* rather than *problem solution* would suggest that they cannot be a silver bullet. They are presented as propositions—rather than hypotheses, recommendations, or conclusions—because the tentative nature of propositions can more readily catalyze transdisciplinary research on the challenges inherent in Middle East policy. Ideally, a dialogue between those researchers and policy actors would spur innovation. Unfortunately, systems thinking can be challenging, requires both discipline and curiosity, and is particularly easy to ignore at the organizational level. The understandable tendency is to fall back on siloed cognitive frames, allowing each individual or organization to consider only their specific role and to ignore the larger, more complex problem and the ways in which culture, institutions, power, incentives, cognition, and personal values and motivations interact (Daniels *et al.*, 2012). However, as the cases and propositions listed above illustrate, a siloed approach is insufficient. Attempts to design and implement foreign policy must necessarily use systems thinking if the goal is to engage other states in their entirety.

Acknowledgements

I would like to acknowledge my professors, factota, and especially my peers at the Telluride Association Summer Program of 2010 in Austin, TX, as my work with them has shaped my views on these issues ever since.

References

Argyris, C. (1962). *Interpersonal Competence and Organizational Effectiveness*, ISBN 9789994174034 (1976).

Argyris, C. (1964). *Integrating the Individual and the Organization*, ISBN 9780471033158.

Argyris, C. (1965). *Organization and Innovation*, Homewood, Il.: R.D. Irwin.

Argyris, C. (1982). *Reasoning, Learning, and Action: Individual and Organizational*, ISBN 9780875895246.

Argyris, C. and Schön, D. (1974). *Theory in Practice: Increasing Professional Effectiveness*, ISBN 9780875892306.

Bánáthy, B.H. (1998). "Evolution guided by design: A systems perspective," *Systems Research and Behavioral Science*, ISSN 1099-1743, 15: 161-172.

Bongars, J. (1611). "Gesta Dei per Francos," trans. in O.J. Thatcher and E.H. McNeal (eds.) (1905), A Source Book for Medieval History, pp. 513-17, http://www.fordham.edu/halsall/source/urban2-5vers.asp.

Checkland, P. (1981). *Systems Thinking, Systems Practice*, ISBN 9780471279112.

Cleveland, W.L. and Bunton, M.P. (2009). *A History of the Modern Middle East*, ISBN 9780813343747.

Corning, P.A. (2002). "The re-emergence of 'emergence': A venerable concept in search of a theory," *Complexity*, ISSN 1099-0526, 7(6): 18-30, http://www.complexsystems.org/publications/pdf/emergence3.pdf.

Daniels S.E., Walker, G.B. and Emborg, J. (2012). "The Unifying Negotiation Framework: A model of policy discourse," *Conflict Resolution Quarterly*, ISSN 1536-5581, 30: 3-32.

Daniels, S.E. and Walker, G.B. (2001). *Working Through Environmental Conflict: The Collaborative Learning Approach*, ISBN 9780275964733.

El-Messiri, S. (1978). *Ibn al-Balad: A Concept of Egyptian Identity*, ISBN 9789004056640.

Fiske, S. and Taylor, S. (2007). *Social Cognition, from Brains to Culture*, ISBN 9780073405520.

Frej, W. and Ramalingam, B. (2011). "Foreign policy and complex adaptive systems: Exploring new paradigms for analysis and action," http://www.santafe.edu/media/workingpapers/11-06-022.pdf.

Ghannam, F. (2012). "Meanings and feelings: Local interpretations of the use of violence in the Egyptian revolution," *American Ethnologist*, ISSN 0094-0496, 39(1): 32-36.

Hofstede, G. (2001). *Culture's Consequences: Comparing Values, Behaviors, Institutions and Organizations across Nations*, ISBN 9780803973244.

Holland, J.H. (2000). *Emergence: From Chaos to Order*, ISBN 9780192862112.

Keen, M., Brown, V.A., and Dyball, R. (2005). *Social Learning in Environmental Management: Building a Sustainable Future*, ISBN 9781844071821.

Makdisi, U. (2000). *The Culture of Sectarianism: Community, History, and Violence in Nineteenth-Century Ottoman Lebanon*, ISBN 9780520218468.

Maturana, H. and Varela, F.J. (1980). *Autopoiesis and Cognition: The Organization of the Living*, ISBN 9789027710161.

Rice, C. (2005). "Remarks of Secretary of State Condoleezza Rice at the American University of Cairo," http://www.arabist.net/blog/2005/6/20/condoleezza-rices-remarks-from-her-cairo-speech-at-auc.html.

Rogan, E.L. (2009). The Arabs, ISBN 9780465020065.

Röling, N. and Wagemakers, M.A.E. (1998). *Facilitating Sustainable Agriculture: Participatory Learning and Adaptive Management in Times of Environmental Uncertainty*, ISBN 9780521581745.

Thompson, L., Neale, M., and Sinaceur, M. (2004). "The evolution of cognition and biases in negotiation research: An examination of cognition, social perception, motivation, and emotion," in M.J. Gelfand and J.M. Brett (eds.), *The Handbook of Negotiation and Culture*, ISBN 9780804745864.

Varela, F., Maturana, H., Uribe, R. (1974). "Autopoiesis: The organization of living systems, its characterization and a model," *Biosystems*, ISSN 0303-2647, 5: 187-196.

Wildavsky, A. (1964). *The Politics of the Budgetary Process*, ISBN 9780673394910 (1984).

Peter O. Daniels is studying Sociology/Anthropology, Public Policy, and Biology at Swarthmore College, Swarthmore, PA. His particular research interests include Middle Eastern studies, the sociology of water, and interdisciplinary approaches to topics in the Middle East.

Applied

Dealing With Violence, Drug Trafficking And Lawless Spaces: Lessons From The Policy Approach In Rio De Janeiro

Kai Lehmann
Institute of International Relations, Universidade de São Paulo (USP), BRA

Until recently, Rio de Janeiro was one of the most violent cities on the planet. Many of Rio's hundreds of shanty towns were controlled by heavily armed drug gangs taking advantage of the absence of the state. However, since 2008, a policy of pacifying some of the city's most strategically important and violent shanty towns through community policing overseen by so-called 'Unidades Policicias Pacificadoras' (Pacifying Police Units, UPPs) has led to a significant reduction in violence. This article argues that this success is down to the fact that this policy treats the issue of violent crime as Complex Adaptive Systems. As a consequence, it seeks to facilitate a process of self-organization balanced between order, flexibility, rules and freedom. The article will show how Complexity has been applied, what benefits it has brought, what problems remain and what broader lessons can be learned from this experience for public policy-makers elsewhere.

Introduction

B razil has the unenviable reputation of being one of the most violent countries in the world. Between 2003 and 2007 more than 240,000 people were murdered in the country, about 27 homicides for every 100,000 people (Carneiro, 2010a).

Within this context, as Richardson and Kirsten (2005) have pointed out, the situation in Rio de Janeiro deserves particular attention. The city witnessed an explosion of violence from the beginning of the 1980s onwards which continued until very recently and, as a major global tourist destination, the host city of the football world-cup final in 2014 and the Olympic games in 2016, the issue of crime has always received more attention in relation to Rio than it has in relation to other cities in Brazil.

The case of Rio de Janeiro, however, is also interesting because of the innovative policies that are being implemented since the end of 2008 by the governor of the state of Rio de Janeiro, Sergio Cabral, in order to at least begin to address this problem, and whose results so far have shown a significant reduction in homicide rates in the city, as official figures from the *Instituto de Segurança Pública* have shown (http://www.isp.rj.gov.br/Conteudo.asp?ident=260).

This paper aims to investigate these initiatives from within the conceptual framework of Complexity, as defined by Cowan, Pines and Metzer (1994). Particular focus here will be given to the installation of so-called 'Unidades Policiais Pacificadores' (UPPs, Pacifying Police Units) in some of Rio's most violent shanty towns. It will be argued that one of the key reasons for the success of these policies is the fact that they, consciously or not, conceive of the problem of violence in Rio de Janeiro as a Complex-Adaptive System, defined by Dooley (1997) as 'a collection of semi-autonomous agents with the freedom to interact in unpredictable ways whose interactions over time and space generate system-wide patterns.' These systems developed through a process of self-organization, defined by Eoyang (2001) as a process where the 'internal dynamics of a system generate system-wide patterns.' As such, instead of trying to 'solve' the problem through a top-down process, the installation of UPPs has been conceived as a way of changing the patterns of self-organization in affected areas, allowing for the *emergence* of new processes of development, responding to the local boundary conditions of each particular area.

Drawing on a mixture of academic analysis and interviews with elite-actors as well as some of those who implement the policy on the ground, the article will, first, outline the broad context within which the policy was developed. Following on, it will subject the policies to a Complexity-inspired analysis. This application will then allow for suggestions about future steps to be taken in order to take the policy further. Finally, the broader lessons of this policy for public policy-making and avenues for future research will be assessed.

The Context: Rio De Janeiro And The Problem Of Violent Crime

As Carneiro (2010a) points out, in 1975, the murder rate in Rio de Janeiro did not reach 15 per 100,000 inhabitants. By 1995, it had reached 64.9, making it one of the most violent cities on the planet. Over the years, the impact of this increase has been profound. The city has innumerable 'favelas', or shanty towns, dominance over which is often disputed between heavily-armed drug gangs and/or militias whose frequent violent confrontations with each other or the police kill hundreds of innocent people every year. According to Moser *et al.* (2005), in 2003, Rio accounted for 19% of all loss of disability-adjusted life years in the country.

There is an extensive literature investigating this phenomenon which points to a number of causes. As Carneiro (2010a) has pointed out, some have to do with simple political incompetence. This incompetence is linked to 'traditional' problems such as corruption and the 'reach' of the state. Within this context, the performance of the military police has received particular attention. Carneiro (2010b) identifies 4 principal problems which have seriously undermined the ability of the police force to effectively fight violent crime in the city. One is internal police corruption. Second, there is a general problem of indiscipline within police ranks 'with orders simply not being followed' (*ibid*: 64). Third, poor training leads to poor operational conduct and, lastly, there simply are not enough policemen to deal with the multitude of issues within often very difficult physical terrain. Linked to these particular institutional problems are deep structural ones of both the Brazilian economy and society at large which are very evident (and visible) in Rio de Janeiro.

This last point refers mainly to the vastly unequal distribution of wealth in the country. Brazil is one of the most unequal countries in the world and Rio, one of the country's wealthiest cities, displays this inequality graphically. The biggest shanty-town in Rio, Rocinha, is separated from a wealthy neighboring community by one road which marks 'a 9-fold difference in employment, a whopping 17-fold difference in income, and a 13-year difference in life expectancy' (Goldstein & Zeidan 2009: 288). As Leu (2008) has stated, there is also a clear physical division between these areas, with most shanty towns lying on hillsides, some of which directly overlook Rio's wealthiest neighborhoods. As will be shown below, this particular feature of Rio de Janeiro has had a significant impact on the policies adopted in the city both in the past and the present.

Yet, as shown by Guimarães, such inequality has been a structural feature of Brazilian society for literally centuries and is the results of 'formal and informal mechanisms which preserve the existing power structures in all its forms' in particular the police, the judicial and political, as well as the educational system (Guimarães, 2008: 16).

These structural factors combined at the start of the 1980s with some specific circumstances to produce a steep increase in violence: 'The combination of a weak state, economic crisis inherited from the military which led to harsh economic restructuring, and the expansion of the drugs trade, led to an increase in violent crime' (Leu, 2008: 3). The combination of increasingly harsh economic conditions led to rapid urbanization, in particular the unplanned expansion of both Rio de Janeiro and São Paulo as people migrated from north to south in search of work. This led to 'favelalization' of city space, manifested in both the growth of existing shanty towns and the creation of new ones, a factor which has been seen as key to violence in Rio in some studies, such as Richardson and Kirsten's (2005).

At the same time there was a massive expansion of the drugs trade in the city which was often organized by armed groups which, taking advantage of the state's weaknesses, installed themselves in favelas and established alternative power structures for their particular areas, underpinned by its own 'laws' and 'code of conduct' (breaks of which were often summarily punished through torture or death), the provision of some basic social services and other functions traditionally regulated by the state, such as television or the internet, as Carneiro (2010b) has shown.

Analyzing Violence In Rio De Janeiro: Complicated Or Complex?

The problem of violence in Rio de Janeiro, then, is certainly 'complicated'. In other words, many different factors have contributed to the rise of violent crime in the city over the last few decades. This is particularly true in this case because the rise in violence also took place during a period of political transformation, with the change from a military dictatorship towards a, by now, reasonably solid democratic system, followed from the second half of the 1990s by a sustained period of economic growth. Within this context, there was a long-lasting and intense debate-one vividly recounted by *Le Monde Diplomatique Brasil* (2009)—amongst the political elite of Rio de Janeiro about the principal objectives of police work and the right way of transforming especially the military police from an institution protecting the state towards one whose priority is the protection of the population.

However, the view of the problem of violence in Rio being 'complicated' had serious consequences for the development and application of policy. These policies can broadly be divided into two approaches: One can be termed 'containment' whilst the other can best be described as 'confrontation'. It will be worth looking briefly at both of these strategies before the application of Complexity and explication of current policy.

'Containment' was basically a policy of limiting police incursions and, therefore, involvement in those areas dominated by drug traffickers. Arguing that a policy of confrontation would lead to the loss of innocent life, the principal objective of the policy was to contain the expansion of the control drug traffickers would exercise within the city to those areas already under their control. Mostly associated with the ex-governor of Rio, Leone Brizola (1983-86 and 1991-94), whose governorship has been analyzed in great detail by Sento-Sé (2002), this policy was part of an attempt to re-frame criminality as a social problem which deserved consideration under the umbrella of re-democratization and human rights. The physical and geographical structure of Rio's shanty towns, according to this argument, significantly increased the risks to civilian lives in case of frequent incursions. As such, non-confrontation preserved the human rights of the general population, just as containment preserved the peace in significant parts of the city.

Yet, this policy had several unintended consequences. According to Carneiro (2010b), the absence of the state allowed the drug-gangs to establish their control over virtually all the big favelas of the greater Rio area which diminished the chances of the police acting effectively. In other words, containment led to a disproportionate growth of the power and influence of the groups being contained by allowing them to establish tightly-knit nets of contact and an infrastructure for the distribution of drugs and arms which greatly enhanced their efficiency whilst reducing costs. This, in conjunction with a huge increase in the *consumption* of illicit drugs in Rio, especially cocaine, increased the power of these groups exponentially. As shown in Carneiro's (2010a) overview, if one adds in police corruption (in itself fuelled by rising drug-generated profits) and an inefficient prison system—which allowed imprisoned drug-gang leaders to continue organizing their respective groups—it soon became obvious that far from containing violence, the policy actually spread the problem.

The almost logical consequence was the emergence of the opposite policy option: armed confrontation with drug gangs. Developed by governor Marcello Alencar (1995-99), the policy created incentives for the military police to kill as many drug dealers as possible, linking pay rises and promotion to the number of kills a policeman could prove. During Alencar's term in office, the number of weapons seized from favelas rose more than 20%, whilst the quantity of drugs seized shot up by over two thirds (Carneiro, 2010b). For a time, this policy seemed to have the desired effect with the number of homicides declining for the first time since the early 1980s. At the same time, the number of innocent civilians killed also rose sharply, somewhat putting into question the overall impact of these policies on violence in the city. Nor did the policy have any appreciable effect on the levels of drug consumption and therefore drug trading. Dead drug dealers were simply replaced by new ones, pointing to another key difficulty confronting policy-makers: the persistence of often abject poverty and lack of economic opportunities which allowed for 'drug dealing' to remain an attractive alternative for many in their attempts to earn money, as shown by Zaverucha (2001).

These two policy approaches competed for several years, depending on the outcome of elections and therefore the occupant of the governors' mansion. In some cases, a combination of the two approaches was applied, confronting drug dealers whilst trying to 'urbanize' particular favelas in order to guarantee the provision of basic social services, as in the case of the program *Favela Bairro*, analyzed in detail by Goldstein & Zeidan (2009).

Yet, why did none of these policies achieve its ultimate aim of consistently breaking the cycle of violence? At this point the application of Complexity will be useful.

The key problem with either policy was that they conceived of the problem confronted as 'complicated' as opposed to 'complex'. In a complicated system 'it is possible to work out solutions and implement them' (Chapman, 2002). In one policy (containment), the solution was the physical containment and concentration of drug dealers in particular areas. This would lead to the rest of the city living in relative peace. The key determinant variables, then, were the geographical location of the problem as well as the group of people being responsible for the problem (drug dealers). In the second policy, the solution was the physical elimination of drug dealers. At the same time, the policy also had a specific focus on particular areas (the favelas), essentially leaving the rest of the city untouched.

There were, hence, significant similarities between the two policy approaches. Both isolated clearly definable variables which could then be altered or, at least, controlled which would lead to the resolution of the problem identified. As such, the policy solutions developed conceived of violence in Rio de Janeiro as a classic linear system, even though the approaches arrived at different conclusions as to what this signifies in policy terms. In other words, either approach was based on 4 basic principles:

- Order: known causes lead to known effects under all circumstances;
- Reductionism: By observing the behavior of its parts, the behavior of a system could be understood clockwork fashion, free of surprises. The whole was the sum of its parts;
- Predictability: Once the behavior of a system is understood, the future course of events can be predicted by application of the appropriate inputs to the model, and;
- Determinism: Processes flow along orderly and predictable lines with clear beginnings and rational ends (adapted from Geyer and Mackintosh with Lehmann (2005: 34).

Yet, the problem of violence in Rio de Janeiro is not merely complicated, but complex. That is to say that the key characteristics of the problem are the following:

- A number of elements or phenomena;
- Emergence and sensitivity to initial conditions. Its development is at best partially predictable;
- Parts of the system are reducible whilst others are not;
- The elements of the system form coherent patterns over time, and;
- The system is open to its environment and therefore capable of adaptation and survival (*ibid*).

This means that 'the relationship between cause and effect is uncertain and there may not be agreement on the fundamental objectives [of any given policy]' (Chapman, 2002: Foreword). Since such systems self-organize 'policies and interventions have unpredictable consequences [whilst complex systems] also have remarkable resilience in the face of efforts to change them' (*ibid*).

It is here that one can find the reasons for the failure of the policies outlined above to deal with the problem of violent crime in Rio de Janeiro. Neither conceived of the system they were dealing with as a self-organizing Complex Adaptive System. Rather, it was thought that through either containment or confrontation the system could be simplified and its development could be controlled and moved in a more desirable direction. Little to no thought was given to the *reactions* either policy would cause, nor to the fact that the areas where these policies were applied were themselves self-organizing Complex Adaptive Systems which *interacted with*—but were not *subordinate* to—other Complex Adaptive Systems such as the city as a whole or the state government. As such, there was no recognition of the *interdependence* which existed between the systems and their *semi-autonomous* nature. In fact, the attempts to separate those areas *with* a problem (the favelas) and those supposedly *without* a problem (the rest of the city) served as a significant barrier to progress, creating a feeling of 'us' and 'them' and allowing for, some would say encouraging, the emergence of distinct identities, as Goldstein and Zeidan (2009) have demonstrated. In the terminology of Eoyang and Yellowthunder (2005), the policies applied accentuated so-called 'containers' ('the police', 'the state', 'the favela'), and emphasized differences (one part of the city has a problem, one has not; one is 'poor', one is not; one counts with the presence of the state and one does not) at the expense of exchanges between the affected population and the 'state', essentially cutting lines of communications and fostering deep mistrust between the different parts and different groups of the city.

Only with the installation of the 'Unidades Policiais Pacificafdoras (Pacifying Police Units, UPPs) has this pattern began to change.

The *Unidades Policiais Pacificadoras*

Sergio Cabral assumed the governorship of Rio de Janeiro at the beginning of 2007 with violent crime one of *the* key issues facing his administration. One member of his government directly involved with the issue of public safety put it thus in an interview: 'Rio had gone through decades of bitty policies. A policy was implemented only to be changed by the next government. We needed to come up with a strategy which would be resilient and which could be embedded sufficiently to make a difference over the long term'. Critically, therefore, the government, from the very start, was conscious of the need to change *patterns* and recognized that such change may take significant amounts of time, critical components of complex systems.

It took almost a year before this strategy was developed: the installation of permanent Pacifying Police Units in some of Rio's most violent shanty towns, staffed by newly-recruited police officers who receive specific training in community policing. The installation of these units would be announced in advance and would subsequently be augmented the application of various social initiatives, such as education, judicial services etc in what was called 'UPPs social', or *Social Pacifying Police Units*, launched in 2010 (see http://www.uppsocial.org/ for more information).

According to the member of the state government interviewed, the basic objective of the installation of the UPPs was to guarantee the right of the population to move freely and enjoy their civil rights in any given part of the city. The installation of a permanent police presence in areas previously controlled by drug traffickers or militias and the announcement of the installation *prior* to doing so was intended to 'minimise the risk of violent confrontations during the initial phase of the operation and, over time, establish the monopoly of force of the state'. The aim, therefore, was to 'de-militarize' the affected areas. In a second phase, the key objective was to establish a relationship of trust between 'the state' and the population. As the commander of one of the units put it: 'Where we are, there was no effective presence of the state for 30 years so clearly there is mistrust. We need to show practical results in order to win that trust'. For this commander, practical results included taking weapons out of the area, allowing for the provision of basic services, as well as the progressive integration of the area with the 'formal' city. To this end, the staffing of these units by *newly-trained* police officers is crucial. According to one member of the strategic police command,

> We needed to change our training and outlook. We are not [in these areas] to confront, we are there to provide security...to serve the community. As such, it is useful to have policemen who are new. It allows us to change patterns and perceptions, though it will take time'.

Having established the basic parameters of the policy, one critical question was the choice of *were* to establish UPPs, bearing in mind the resource intensiveness of keeping a fixed 24 hour police presence in a particular area and the fact that there are hundreds of shanty towns in the city. According to one member of the state government involved in the development of the policy,

> We had to make strategic choices. The first area to be pacified (the shanty town Santa Marta) was a relatively small community but was a key area for the distribution of drugs and, crucially, weapons [to other regions]. So, it was crucial to [get hold of this problem]. [Since then] the choices have been made on a case-by-case basis and the policy has developed from there...We now have a "waiting list" of places wanting a UPP [so there has been] some cultural change but [all of this] takes time....The key was to break with the cycle of violence in some of these areas and create [an expectation] that the state will have control over areas [previously dominated by traffickers] and that [the state] will stay.

Until the end of 2011, 18 such police units have been established, covering 68 shanty towns, directly benefitting 315,000 people with more than 1 million benefitting indirectly in neighboring districts, the aim being the establishment of 40 such units until 2014 (see http://www.bbc.co.uk/portuguese/noticias/2011/12/111219_qa_upps_jc.shtml for the figures). During the 3 years of the policy, the police has established a presence in some of Rio's most notorious shanty towns, including Latin America's biggest, Rocinha, as well as Rio's most violent complex, the Complexo do Alemão.

The impact of this policy has been felt on many levels. In the most practical terms, there has been a significant reduction in violent crime in the areas affected and in the city as a whole. For instance, according to the statistics collated by the Institute of Public Safety of Rio de Janeiro, the number of homicides in the city declined from 2,155 in 2009 to 1,422 in 2011. In areas close to UPPs, the reduction in crime is even more pronounced, with, for instance, Botafogo, the district which contains the first UPP, experiencing a 71% drop in robberies between 2009 and 2010, as the *Instituto de Segurança Pública* has shown (http://www.isp.rj.gov.br/resumoaisp.asp).

However, critically for the long-term success of the policy, those delivering the policy have perceived a notable change in attitude towards the police: 'We are seen in a different light here. People used to [associate the police] with violence, now they perceive us to be here to help and to *stop* violence', as one unit commander put it. Studies done by the Brazilian Institute of Social Research (IBPS) underscores this impression, with 86% of people saying UPPs had made their area 'much better' or 'better' whilst 79% said that the presence of the police had eliminated the presence of armed gangs in their area. This led to 80% of respondents saying that the image of the military police had become 'much better' or 'better' (IBPS, 2010).

According to the member of the state government, this transformation in the image of the police, whilst incomplete, was what *allowed* for the development and expansion of the policy: 'It is critical. There is now an *expectation* that, eventually, we will have a police presence in [all places] which is a huge transformation. It puts pressure on us, but it shows that we have had an impact'.

What, though, is 'complex' about this policy and can Complexity be used to explain this success? The first key decision was the definition of the objectives: 'We do *not* intend to stop all trafficking of drugs, nor all criminality. Our aim is to break the pattern of violence which persisted in these areas and allow for a different pattern [of development]', according to the member of the state government. It was for this reason that the establishment of UPPs has always been announced prior to their establishment, as the government member explained:

> We recognise that this will mean that some drug dealers will escape. However, outside their communities they do not have the same influence and power as they did inside. We also wanted to consciously avoid armed conflict, to show that [peace] is the aim, as opposed to killing. [We have] received criticism for that, but the important thing was to change the pattern.

In fact, the government member pointed out, one of the reasons for the delay in initiating the policy was the internal debate about this approach: 'Some just wanted more resources for the police so that more [drug dealers] could be killed...it took a while to convince [these people] that we needed a different approach.'

It was also recognized that the pattern was different for each shanty-town, as the government member continued: 'We studied each case exhaustively before deciding on whether to go in, with how many men and what our priorities there would be. It is critical that we know each place and can respond to its particular needs'. In other words, there was recognition of the particularities of each case, be it in geographic location, size, the influence formerly exercised by alternative power structures etc.

Thirdly, there was a conscious engagement with the local population. The commander of one of the units put it this way:

> [The population] know what they need much better [than we do]. We need to respond to their wishes otherwise they will not [trust us]. There needs to be constant feedback and we have consciously engaged with [representatives of the community] to show than that we are here for them.

As such, the implementation of the policy is essentially a *social* activity, with each local area having significant autonomy to decide its particular needs, a key plank for any successful process of self-organization, as Rihani (2002) has shown.

Critically, therefore, the *basis* upon which the policy was developed was different which, in turn, influenced the particular phases which followed. It was recognized that violent crime in Rio de Janeiro was a societal *pattern* rather than a question of particular people or clearly definable variables. The *strategic* decision therefore was to try and change this pattern by changing the local boundary conditions of those areas which had a critical role in sustaining this pattern. Crucially, however, there was recognition that these strategic areas confronted often very different local boundary conditions which made engagement with the local population and, hence, de-centralisation a necessity. This informed the *operational* policy with its focus on local engagement and relative operational autonomy for the unit commander.

As such, one can therefore identify the three essential characteristics for any process of self-organization, as outlined by Eoyang and Yellowthunder (2005). The UPPs provide the key 'container' for the process of self-organization. In other words, they provide the stable order without which no coherent development can take place. In addition, there is a commitment from the state government that no area of the city should be 'off limits' for the state and that the presence in any territory has to be permanent. The operational autonomy given to unit commanders is essential to respond to the differences which exist across time and space and ensures that they can enact measures most appropriate for their particular unit is a critical strategic decision. However, this process of de-centralisation can only work if there is engagement with the respective local community. In other words, there must be the possibility of exchange and engagement so that these critical differences can be expressed. This applies not only to the pacified community itself but also to its immediate neighborhood: 'We know there

E:CO Vol. 14 No. 4 2012 pp. 51-66

are key differences between favelas and the rest but we need to make sure that we see them as part of one city', according to the member of the government. 'We cannot separate what happens in the favela from the rest.'

Crucially, therefore, the state government, this time, did *not* make an attempt to simplify the policy-landscape. Rather, recognizing this complexity led to crucial strategic decisions (for instance, the need for de-centralisation) which were then followed up by crucial operational decisions, such as operational autonomy for unit commanders. In other words, there was a clear and consistent thread informing policy development *and* implementation, as well as awareness of the imperfect results the policy will produce. The consequence has been a reasonably successful policy in that it has changed *patterns* of development and *expectations*: 'People want us here now and they feel safer', according to one unit commander. According to him, this expectation of security and peace has led to new patterns of behavior and a desire on the part of some members of the old drug gangs to return to the formal labour market. One spin-off effect been the 'legalization' of lots of other services that people need but often did not get from the state. The same commander states:

> People are actually beginning to demand that their electricity be provided by the local electricity company, that their water be provided legally, that their satellite TV be provided by a satellite TV provider. They like the fact that they have a fixed date to pay their bills rather than when a trafficker knocks on the door. It allows them to plan and shows them that the state is a good thing not a threat.

Within the context of Complexity as a conceptual framework, the above points are critical. They clearly point to the necessity of *both* general rules *and* local variety. Amongst the general rules established by this policy are: the indivisibility of city space, the monopoly of the state as the guarantor of security and other basic services, the presence of the agents of the state in all parts of the city and, therefore, the guarantee of the same constitutional rights for all parts of the population. *At the same time*, there is recognition of local variety. Within this context, success signifies new patterns of self-organization, as opposed to 'solving' the problem of drug consumption and crime. The government member agrees: 'I think we are changing expectations and that is a good thing. However, this creates its own problems and [shows] that we need to keep working and adjusting'.

None of this means that the process is completed or that there aren't considerable problems still to be overcome. Despite the undeniable progress, Rio is still an extremely violent city with shocking levels of poverty, corruption, a thriving illicit drugs market and some areas dominated by alternative power structures, with militias made up of corrupt policemen and other elements dominating significant areas in the west of the city, as Freitas (2010) has shown. As such, the underlying causes for the explosion of violence from 30 years ago have not been eradicated. Therefore, political continuity is crucial in order to embed allow for the changes of patterns outlined above to be embedded. However, as many others have pointed out, these problems go way beyond Rio and even way beyond particular policy areas. Mistrust of the state, corruption, as well as a general disrespect for life is part of Brazilian culture and will take decades to change, as Guimarães (2008) has explained.

At the same time there are particular problems brought about by the policy. One has to do with costs. UPPs are incredibly resource intensive, concentrating a relatively large number of policemen in a relatively small area. This, in turn, could have a serious impact on the fragile change of the relationship between the police and the population: What if the expectations of the population cannot be fulfilled? What if not every shanty town which would like an UPP can get one? What about richer parts of the city? What if some of the UPPs installed do not have the same success as the first ones? All these questions are, as yet, unanswered. Again, constant vigilance is required here. For instance, there needs to be complete rigor in punishing misbehavior of police officers so as to not undermine the early success, something which historically has not always happened, as the member of the current state government readily acknowledged: 'We need to be strict and consistent and, historically, that has not always been the case.'

The very success of the policy is already starting to have unintended consequences: Firstly, both the strategic and the operational commander interviewed for this research have alerted to the problem of police overload in the respective areas:

> We are being asked to do things that we cannot do, such as sorting out
> problems with light and the like, for which we are not responsible. We are
> asked to sort out problems with rent and many other things. The worry is that
> people will see us as substitute for other important groups, such as the resident
> associations, thereby weakening the presence of [civil society].

according to the unit commander. 'There needs to be education about how the state works and who does what', according to the strategic commander.

A second practical issue, according to one of the unit commanders, has been the growth of the pacified shanty towns as people move in search of security: 'We have people moving here from all over Rio as well as [from out of state].' This has led to problems in terms of space, access to services and, as O Globo showed in its edition of 30th May 2010, increasingly, sky-rocketing property prices and rents, a key issue for a community where the vast majority of residents are poor. As such, several other initiatives will be needed to deal with the new environment created by the success of UPPs.

E:CO Vol. 14 No. 4 2012 pp. 51-66

Implications And Conclusions

The above list of problems is in no way exhaustive and is intended merely as an illustration of the *kinds* of issues which have been left either unresolved or have arisen as a result of the policy itself, illustrating the ongoing nature of the process of self-organization.

Despite these problems, however, the overall assessment of the first three years of this policy is overwhelmingly positive. Without doubt, the government has changed patterns of development in some of the most challenging areas of Rio de Janeiro and has, in the process, created a different perception of at least parts of the state and different expectations by the population in relation to at least some policy areas. This represents significant progress.

In terms of the general implications of the policy approach demonstrated by the UPPs, it is striking how they take account of some of the key principles of Complexity thinking: the focus on patterns, the type of objectives being formulated, the importance of engagement and de-centralisation to make changing these patterns a social activity. However, equally striking was the fact that none of the policy-makers and practitioners interviewed for this research ever made specific reference to Complexity as a conceptual framework to inform their policy. In fact, asked about this, all interviewees displayed a certain apprehension to get drawn into what they considered to be theoretical debates: 'I need to deal with facts on the ground. I needed to change [these facts], so that was our starting point', according to the strategic police commander. For him, decentralizing, opening channels of communications, being aware of differences across time and space were *logical* steps to be taken in response to the relative failure of previous policies which did *not* take account of these factors. For advocates of Complexity, therefore, one of the key lessons to be drawn in their attempts to insert themselves more and more into public policy thinking is to emphasise these concepts in relation to 'common sense policies' rather than as part of an elaborate theoretical debate. Complexity, as presented in this article, is a conceptual framework to inform the development and analysis of practical policies rather than a fully-coherent theory.

A second key lesson to be drawn is the need to be flexible when applying Complexity concepts to different areas of public policy and to be *aware* of the specific circumstances within which one is acting. In the case of Rio, for example, this means being aware, on the one hand, of the deep and mutual distrust between the state and the population whilst, on the other, being conscious of the strong paternalistic culture which exists, which, in this case, translates into an expectations that it is the authorities (whoever they may be) to make and apply the rules (on this culture, see Giambiagi, 2007). As one unit commander pointed out, it is extremely difficult to reach a point where residents *also* feel responsible for their area: 'This will take years....[the population] has huge expectations and many ideas but little notion that they, too, can contribute to [realizing these ideas].'

As such, local circumstances and culture have a huge impact on how and to what extent the basic principles of Complexity are applied. Whilst, for instance, it is important to de-centralise in order to take account of local boundary conditions, such a process is different in Brazil than it would be in, say, Germany where the relationship between the population and authority exists within a different context.

This leads to a third lesson: De-centralisation, flexibility and, therefore, the granting of a degree of autonomy to local actors does *not* imply a lessening of the importance of the state. As has been shown by the above case, changing patterns of self-organization *depends* on the existence of a stable framework of rules, adherence to which also *requires* the state. Balancing authority and autonomy, rules and freedom, rights and responsibility is a key task for any government and, as recently shown by Friedman and Mandelbaum (2011) in relation to the United States. The *exact* balance between these will depend on local boundary conditions. Therefore, finding this balance is often a case of trial and error, which emphasizes again the need for flexibility.

As such, one key task for Complexity practitioners now is to apply these principles to many different policy areas and political issues. *Raising awareness* of the fact that adhering to Complexity-principles is not a *threat* to the state but merely suggests a different *role* for it is a key task. The above case shows that some policy-makers 'get' Complexity inherently in some cases. Yet, bearing in mind the uncertainty of outcome, one key area for further work is the question of how advocates of Complexity can persuade policy-makers to *consistently* adopt models and methods inspired by Complexity. In other words, work needs to be done to *legitimise* the concepts of Complexity as guides to public policy. Here further engagement with the policy-making community is critical in order to show that what is being done, in this case in Rio de Janeiro, *does* have sound methodological and epistemological foundations which ought to be applied more broadly.

References

Carneiro, L.P. (2010a). "Mercados ilícitos, crime e segurança pública: Temas emergentes na política brasileira," *CLP Papers*, 5(July).

Carneiro, L.P. (2010b). "Mudanías de guarda: As agendas da seguranía píblica no Rio de Janeiro," *Revista Brasileira de Seguranía Píblica*, ISSN 1981-1659, 4(7): 48-71.

Chapman, J. (2002). System Failure: Why Governments Must Learn to Think Differently, ISBN 9781841801230.

Cowan, G.A., Pines, D. and Meltzer, D. (1994). *Complexity: Metaphors, Models and Reality*, ISBN 9780738202327.

Dooley, K. (1997). "A complex adaptive systems model of organization change," *Nonlinear Dynamics, Psychology, and Life Sciences*, ISSN 1090-0578, 1(1): 69-97.

Eoyang, G. (2001). *Conditions for Self-Organizing in Human Systems*, unpublished Doctoral Dissertation, Cincinnati, OH: The Union Institute and University.

Eoyang, G. and Yellowthunder, L. (2005). "Beyond bureaucratic boundaries: A case study in human system dynamics," paper presented at the *Complexity, Science & Society* Conference, University of Liverpool, September 2005.

Freitas, C. (2010). "Depois dos traficantes, o desafio de acabar com as milícias," Veja, ISSN 0100-7122, 28th November.

Friedman, T.L. and Mandelbaum, M. (2011). *That Used to Be Us: How America Fell Behind in the World and How We Can Come Back*, ISBN 9781250013729.

Geyer, R. and Mackintosh, A. with K. Lehmann (2005*). Integrating UK and European Social Policy-The Complexity of Europeanisation*, ISBN 9781857757644.

Giambiagi, F. (2007*). Brasilí, Raizes do Atraso: Paternalismo versus Produtividade: As dez Vacas Sagradas que Acorrentam o País*, ISBN 9788535224412.

Goldstein, J. and Zeidan, R.M. (2009). "Social networks and urban poverty reduction: A critical assessment of programs in Brazil and the United States with recommendations for the future," in J. Goldstein, J.K. Hazy, and J. Silberstang, J. (eds.), *Complexity Science and Social Entrepreneurship: Adding Social Value through Systems Thinking*, ISBN 9780984216406.

Guimaríes, S.P (2008). *Desafios Brasileiros Na Era dos Gigantes*, ISBN 9788585910792.

IBPS (2010). "Pesquisa Sobre a Percepíío Acerca das Unidades de Polícia Pacificadora," PR 004-10-UPP-25.01, Rio de Janeiro: Instituto Brasileiro de Pesquisa Social.

Le Monde Diplomatique Brasil (2009). "Nío matarís," 4(37).

Leu, L. (2008). "Drug traffickers and the contestation of city space in Rio de Janeiro," E-Compís, ISSN 1808-2599, 11(1): 1-16.

Moser, C., Winton, A. and Moser, A. (2005). "Violence, fear and security among the urban poor in Latin America," in M. Fay (ed.), *The Urban Poor in Latin America*, ISBN 9780821360699.

O Globo (2010). "Imíveis em favelas com UPP sobem atí 400%," 30th May.

Richardson, L. and Kirsten, A. (2005). "Armed violence and poverty in Brazil: A case study of Rio de Janeiro and assessment of Viva Rio for the Armed Violence and Poverty Initiative," report by the Centre for International Cooperation and Security, Bradford: University of Bradford.

Rihani, S. (2002). *Complex Systems Theory and Development Practice: Understanding Nonlinear Realities*, ISBN 9781842770467.

Sento-Sí, Joío T. (2002). "O discurso brizolista e a cultura política carioca," *Varia Historia*, ISSN 0104-8775, 28: 85-104.

Zaverucha J. (2001). "Poder militar: Entre o autoritarismo e a democracia," *Sío Paulo em Perspectiva*, ISSN 0102-8839, 15(4): 76-83.

Kai Enno Lehmann, PhD (2010), University of Liverpool, UK, is currently a Lecturer ('Professor Doutor') at the Institute of International Relations, University of São Paulo (USP). Previously worked at the Pontifícia Universidade Católica (PUC) in Rio de Janeiro (2007-2011) and the University of Liverpool, United Kingdom (2002-2007). Principal research interests include: Application of Complexity to security questions, foreign policy decision-making processes and European integration.

Applied

A Policy Paradox: Social Complexity Emergence Around An Ordered Science Attractor

Alice E. MacGillivray[1] *& Krista G. Gallagher*[2]
1 Institute for Social Innovation, Fielding Graduate University, CAN
2 Royal Roads University, CAN

This paper explores the question of whether people involved with a successful watershed policy initiative embraced and/or negated the complexity with which they worked. The setting was Lake Simcoe, in central Canada: an area important for fisheries, agriculture, tourism, recreation and citizens' identities. Human activities had impacted water quality, and planned development posed further threats. Although government had supported considerable scientific data collection, citizens became frustrated by what they saw as a lack of regulatory and enforcement work. Citizens embarked on a range of creative pressure tactics for change. In early stages, citizens felt marginalized, but over time they were included in increasingly meaningful ways. This paper explores several complex system themes in interview transcripts, including initial starting conditions, attractors, and boundaries. A key finding is that citizens used scientific data as an attractor to enable their inclusion for a more complex range of agendas and benefits.

Introduction

This paper presents a successful policy initiative called the *Lake Simcoe Protection Plan*, intended to protect the Lake Simcoe watershed in Ontario, Canada. Using complexity theory lenses, it explores the question of whether public servants or citizens involved with the initiative embraced and/or negated the complexity with which they were working.

Participants were involved with the process as public servants, local business owners, non-governmental group members, and youth workshop participants. Their experiences were probed through semi-structured interviews. Participants were not asked directly about systems or complexity theory concepts. In order to shed light on future policy work, the authors analyzed the 14 interview transcripts to explore connections—or lack thereof—with complexity theory.

The Research Context

The editors of this issue have acknowledged that the complexity of modern policy work demands more than an understanding of challenges and opportunities. It requires a comprehensive understanding of human and social processes, especially in parts of the world where elected officials espouse democratic principles and practices. A combination of analytical and public knowledge holds promise for better coherence amongst citizen expectations, scientific goals and political achievements. Yet it is rare for scientists and citizens to speak a common language, or even find forums where both voices are validated, as was the case with the Lake Simcoe initiative.

In this paper, we use the term *Lake Simcoe Protection Plan*—or simply *Plan*—as the umbrella term for the *Lake Simcoe Protection Act*, policy process and their outcomes.

We explored this case from a complexity perspective because 1) multi-stakeholder policy work is notoriously difficult and complex; 2) the approach to development of this plan was innovative and 3) the process was considered successful. Its distinctiveness and importance are summarized below:

- Scarcity of clean, fresh water is a global issue; Lake Simcoe is a significant water body in one of the most water-rich countries in the world.

- The Plan is Ontario's first water policy that encompasses an entire watershed.

- The Plan built on rich scientific data from 30 years of monitoring through the Lake Simcoe Environmental Management Strategy (LSEMS).

- The Plan's creation involved pressure from and robust dialogue with a variety of non-governmental participants, including citizens and First Nations communities.

- Citizen voices were integral in the creation of policies, mandates and desired outcomes.

Background On The Policy Initiative

Beginning in the 1970s, the Lake Simcoe Conservation Authority and the provincial government monitored the health of the lake through LSEMS. However, citizens were increasingly concerned about the lack of enforceable mechanisms to protect the lake, especially with increased shoreline development. Citizen groups raised awareness through campaigns, outreach events, and meetings. They lobbied the Ontario government for legislation. In 2007, the Premier of Ontario, Dalton McGuinty, arrived at a rally for Lake Simcoe and announced that the government would create a Lake Simcoe Protection Act. This Act was passed in 2008. The Lake Simcoe Protection Plan was developed in 2009 for "implementing projects, highlighting targets of phosphorous reduction, putting forward limitations on new development and growth, and…voluntary and mandatory measures for the province and municipalities to protect Lake Simcoe" (Gallagher, 2012: 9).

Policy Context

Politically, Canada operates under the Westminster model of Parliament, without clear separation between the parts of government that create and implement laws. Some consider this model "a major achievement in social order and stability" reflecting a modernist or ordered view of reality (Geyer & Rihani, 2010: 23). Federal and provincial levels of government have ministries led by elected officials with titles such as *Minister of the Environment*. Canada's provincial governments are responsible for most land-use decisions.

Evidence-based policy training—which can appear to conflict with complexity thinking--has become common. However, Warburton and Warburton (2004) note there may not be enough evidence, enough kinds of evidence, or the evidence may not complement important agendas such as Canada-U.S. relations (where water access can be controversial). Zussman (2003) describes one such mismatch from health policy: "Evidence also shows that cigarettes kill more people annually than does marijuana…What would be the impact of legalizing the possession of marijuana, for example, on our ability to cross the Canada-United States border at a time when the United States government has a no-tolerance attitude toward drug use?" Head (2008) describes shortcomings of evidence-based policy work for complex problems, including the many bases of evidence valued by different groups. Members of an Indigenous community might—for example—be skeptical when findings of a scientific study funded by the dominant culture appear to conflict with their experiential learning. Gerrits critiques anthropocentric and modernist undercurrents of public policy work. He notes the frequency with which it is "assumed that the decision-maker is in full control of the physical system" (2010: 19).

Scholars acknowledge the complexity of modern policy work (Dennard *et al.*, 2008; Pawson *et al.*, 2005). Runhaar *et al.*, write that policy workers who deal with sustainability must be able to evaluate different policy approaches for multi-actor contexts and "produce knowledge that is scientifically valid, relevant to the policy debate, and accepted by stakeholders" (2005: 15). Pawson *et al.* point out that policy work deals with "complex social interventions which act on complex social systems," and propose what they call a realist review process to "combine theoretical understanding and empirical evidence, and focus on explaining the relationship between the context in which the intervention is applied, the mechanisms by which it works and the outcomes which are produced. The aim is to enable decision-makers to reach a deeper understanding of the intervention and how it can be made to work most effectively" (2005: S1-21).

Challenges to evidence-based policy approaches relate not only to the nature of knowledge, but also to the importance of interactions in complex systems. Evidence-based policy grew out of evidence-based medicine, and is rooted in assumptions about rigorously defendable and transferable knowledge. Such assumptions become problematic as several systems come into play. Boulton (2010: 35) states that interactions are downplayed in the "prevailing mechanical, scientific worldview…Complexity thinking emphasizes that it is rare to be able to ignore the systemic nature of the world in which we live. And, of course, to connect policy makers and policy making is to challenge existing governance structures and power bases."

These studies suggest the complexity of policy work is increasingly acknowledged, but the comfort of more ordered models remains alluring in the potentially tempestuous world of politics. To address this dilemma as *either/or* would negate lessons from complexity science. Geyer and Rihani (2010: 29) state that "orderly and disorderly frameworks are equally flawed" and that complexity theory "acts like a synthesis or bridge between these two."

Methodology

Overall

This study is qualitative and exploratory. It incorporates elements of grounded theory and phenomenography. Analyzed data came from semi-structured interviews with people who participated in the creation phase of the Plan. Some of the analysis done for the second author's thesis were used in this paper, and are referred to as the "initial research." The initial research examined environmental policy creation, and focused heavily on the success of the multi-stakeholder involvement in the creation phase of the Lake Simcoe Protection Plan. New analysis was conducted for this paper from a complexity perspective.

Participants

In the 2010-2011 initial research through Royal Roads University, 16 participants were identified through Plan literature and were contacted by e-mail. Participants included actors from government, youth workshops, local businesses, interest groups, and non-governmental groups. Initially, 8 of the 16 responded and after a prompt, 3 more agreed. Through interviews, 7 other individuals were identified and 3 of them participated, resulting in a total of 14 interviews. No one explicitly refused to participate.

In the initial research, non-governmental groups were compared with government groups for two reasons: 1) government actors have a position of power and can strongly influence legislation creation, and 2) non-governmental individuals are typically motivated by factors outside of their professions. This categorization became helpful for analysis through a complexity lens as well.

Themes For New Analysis

The authors coded text with four pre-determined [complex] system-related themes:

1. Systems explicitly referenced;
2. Boundaries, with no preconceived notions of type;
3. Initial starting conditions, and;
4. Degree of predictability, including related strategies such as detailed planning vs. probes and observation.

Coding And Analysis

Interviews were transcribed verbatim. The text was first coded by participant and associated demographics. Color-coding was used to identify themes, and memos were used to note whether statements were coherent with a complexity perspective. For example, in the fourth theme, a participant might speak about the importance of detailed prediction and step-by-step planning, or describe limits of ordered methods in a complex and unpredictable environment. Phenomenography informed this exploration of variation in perspectives. A table was developed to show which participants spoke to each theme.

Findings

This section describes findings relating to the four themes above, as well as relevant findings from inductive analysis during the initial research.

Different participant groups are compared, but these categorizations are somewhat artificial. There are degrees of connectivity, where "in a social context, each individual belongs to many groups and different contexts and his/her contribution in each context depends partly on the other individuals within that group and the way they related to the individual in question" (Mitleton-Kelly, 2003: 6). In the Lake Simcoe process, some individuals were citizens as well as conservation authority employees. Some were cottagers on Lake Simcoe and also worked for environmental non-governmental organizations (ENGOs). Such overlapping and potentially conflicting roles added to the complexity.

Systems Referenced By Participants

Policy work is becoming more complex and decision makers need to consider "ways to incorporate a balance among economic, ecological, social, and cultural value creation into their business models" (Porter & Derry, 2012: 33).

This coding pass explored systems referenced by participants. Participants were not asked directly about systems and they rarely used systems concepts explicitly (although one participant did reference feedback loops). They all spoke about systems they cared for, or that needed attention in this policy creation process.

Citizen participants spoke most frequently about social, cultural and relational systems. They were also cognizant of the ecological systems that were at risk because of declining water quality. Residents' identities were often tied to their homes on the lake. Citizen concerns stemmed from an intersection of ecological, social and cultural systems; they saw a complex collection of problems, which prompted them to act. In a sense, threats in the watershed were threats to their identities, which drove the passion behind many of their lobbying efforts to create an Act.

Government actors emphasized political and economic systems. Fisheries were in decline, and tourism was impacted by poor water quality. These issues, beach closures and questions about large development projects near Lake Simcoe—which could fuel the economy while damaging natural systems—were becoming hot issues in nearby municipalities. Participant 1 (P1) noted: "the fishery was key—it was being impacted, and was going down. There is a huge economic interest in the fishery. It wasn't self-sustaining anymore, this got played up where it was almost viewed as an endangered species. On a local scale, this is not what you want happening in your watershed. This was a tipping point, it had to be now or never and gave a sense of urgency." The intersecting economic and ecological concerns of a declining fishery exemplify the intersecting systems involved in the creation phase of the Plan.

Intersections of the citizen and government groups created spaces for learning, and enabled the creation of two multi-stakeholder committees to inform policy makers. Diverse perspectives and rich interactions nourished the creation of the Plan. Mitleton-Kelly notes that "it is more natural, or at least less ambiguous, to speak of *complex behavior* rather than complex systems. The study of such behavior will reveal certain common characteristics among different classes of systems and will allow us to arrive at a proper understanding of complexity." (Mitleton-Kelly, 2003: 4). In the Lake Simcoe case, numerous complex behaviors illuminated an underlying shared concern about the health of Lake Simcoe. As P3 stated "even development and industry realized that the water needed to be improved." This common concern became a touchstone for diverse groups.

Boundaries

We explored participants' references to boundaries because the *boundary* concept is central to [complex] systems thinking and related decision-making (Midgley, 2000). Coding under the umbrella of boundaries was done without any preconceived notions of what the boundary-related sub-categories might be.

Administrative And Role-Related

Twelve of the fourteen participants spoke about boundaries related to roles, administrative units and jurisdictions. Most of those comments related to pressure from one or more groups shifting the ways in which other groups worked. Government employees frequently said things such as "there was a lot of pressure from citizen groups to be heard and make steps towards more effective protection of Lake Simcoe" (P7). They also spoke to cross-jurisdictional challenges. Citizens described how they had become involved, learned about lobbying, put pressure on government, and pushed government until they responded.

Disciplinary Boundaries

Many participants described work with an unusually broad range of conservation-related departments. Comments about the importance of the watershed were more comprehensive, and included tourism, agriculture, personal enjoyment and art. Several spoke about challenges of reaching consensus across boundaries. P5 described the artificiality of disciplinary boundaries: "Everything is in silos: environment, art, science, music, recreation; whereas in reality, these things are not separate." Ruddick (1996: 132) describes this shift from systems as separate and distinct to intersecting, moving "scholarship away from arid, endless debates that attempted to identify which system was predominant. Public servants chose to sweep in a broader-than-typical range of government voices, yet—based on interview data—emphasize certain fields. It is not clear from the interviews whether these boundary choices were subconscious, designed for efficiency, or politically motivated. One person did mention a broad array of interested ministries within government, which suggests some were present without holding equal status.

Edge-Effect™

Edge-effect is adapted from its ecological roots: one discipline behind the development of complexity science. The edges of intersecting habitats are often rich places for biodiversity. Estuaries are places where land, fresh water and saltwater habitats collide and are much more diverse and productive than the sum of the contributing habitats. The first author has developed the concept of intellectual estuary™ where people from different experiential bases come together. P2 was among those who described powerful intersections: "There's a symbiotic relationship between MOE [Ministry of Environment] and Campaign Lake Simcoe. The MOE can't be political and even if they recognize that there is more funding needed, they can't ask for more funding. They can't be activists or advocates."

Boundary Choices

Systems thinker C. West Churchman (1998) wrote about the impacts of boundary choices we make daily. People rarely reflect on the ethical implications of those choices, or even make them consciously. P7 was among those who mentioned boundary choices, saying that the plan's "main objectives are en-

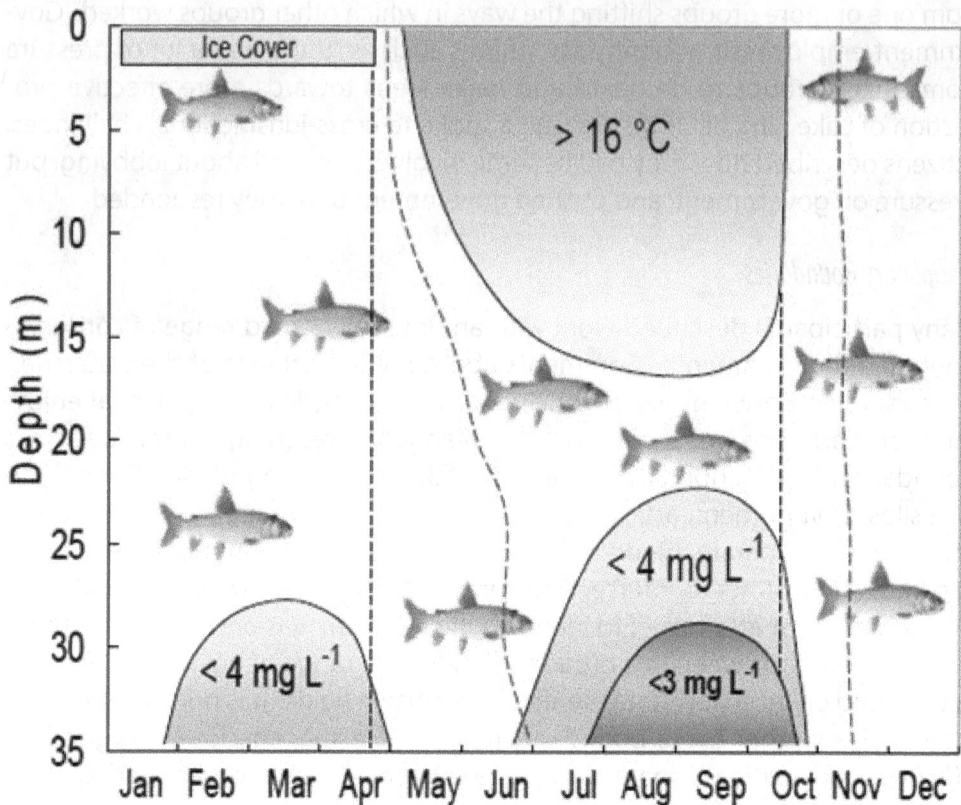

Figure 1 *Illustration of the Habitat Squeeze Experienced by Lake Trout in Lake Simcoe During the 1980s and Early 1990s. Illustration from 2003 LSEMS Report*

vironmental". Communication from government websites often reflected this boundary choice.

P7 spoke about benefits of moving beyond the lake to a larger watershed boundary, and how he discouraged those who wanted to further broaden boundaries in ways that would have been too overwhelming. Another participant said it was helpful to have a strong focus on Lake Simcoe, even though "every watershed in the province needs help."

Citizen groups pushed boundaries through initiatives such as the Ladies of the Lake Calendar. They decided to do something "completely off the wall" and discovered the calendar changed how people thought about the lake, raised one quarter million dollars, and helped citizens move from concern to a sense of responsibility and action (P5).

Boundary Critique

In Systemic Intervention (2000), Midgley describes tensions and rituals that develop when a powerful group sits at the core, with others outside the core boundary on the margins. He presents this as the *theory of boundary critique*.

Figure 2 *Image from the 2009 Ladies of the Lake Calendar. Photo by Jim Panou. Reproduced with permission by the Ladies of the Lake Conservation Association.* www.lakeladies.ca.

For the Plan, government formed the core. Core groups typically have well-developed cultures, which shape the knowledge that is valued and how communication takes place. The core can value or devalue a marginalized group. Midgley uses anthropologist Mary Douglas's terms *sacred* and *profane*, for these value-based categories.

Early in the creation phase, citizen groups felt neglected and had to push hard to be heard. Public servants acknowledged the value of multiple perspectives, but had to overcome inertia. One citizen noted: "it seems like that was the dark ages… One of the principles of the LSPA is that the Minister responsible for the Act, would be advised from a committee made up of government and private sector" (P5). A government participant (P7) reflected: "I've been with MOE for 11 yrs and I've never seen anything like this, it's really inspiring and it makes being part of it so exciting you actually get to talk to people on the ground who care and are interested." Using the language of boundary critique, citizens were by definition outside the *core*, but their involvement became welcomed and could be labeled as sacred. Some citizens mused as to whether this [sacred] status was sustainable.

Initial Starting Conditions

Most participants emphasized LSEMS as a key foundation for data and communication. P4 noted that "[The Plan] was built over time from LSEMS, this was a partnership strategy together with the municipalities. LSEMS included stakeholders as well, that process had been going on for 15 years, there was lots of work already done, a good framework to build on." Government groups were pushed by citizens; both agreed it was the right time to create the Plan. Mitleton-Kelly calls this coevolution in complexity theory. This is where the *"evolution of one domain or entity is partially dependent on the evolution of other related domains or entities; or that one domain or entity changes in the context of others…* In human systems, coevolution in the sense of the *evolution of interactions* places emphasis on the relationship between the coevolving entities" (Mitleton-Kelly, 2003: 7). Progress towards an Act was achieved through such coevolution. Such processes can "change the perspective and the assumptions that underlie much traditional management and systems theories" (Mitleton-Kelly, 2003: 7). Without LSEMS the evolution and policy work would have been much different.

Conn's research complements Midgley's by describing essential differences between *vertical* (government in core) and *horizontal* (citizens in margins) spheres. "The horizontal sphere cannot simply be a replication of how the vertical sphere operates, but rather engages [its own] methods and strengths." Coevolving with—rather than adapting to—other stakeholders is important as "often civic engagement and inclusion tend to be on terms familiar and suitable to the *vertical hierarchical* [core government] dynamic" (2010: 9). This hierarchical approach changed over time in the Lake Simcoe context. Government actors included the horizontal sphere through dialogue, workshops, and other forums, which revealed diverse interests and strengths (Gallagher, 2012).

Degree Of Predictability: Detailed Planning Vs. Probes And Observation

There was considerable uncertainty during the Plan's creation. Citizens did not know what would happen if they put their time, finances, and livelihoods on hold while they fought for the legislation. Government actors could not predict outcomes of a strong intra-governmental group and intense pairing with citizen and non-governmental groups.

Complexity theory tells us that "the search for a single 'optimum' strategy may neither be possible nor desirable" (Mitleton-Kelly, 2003: 14). Although government participants described a relatively ordered approach, they lowered their resistance to unordered and creative approaches used by non-governmental groups. They were able to progress through small adaptations and acknowledgements of the 'other'. P1 noted "I was involved in taking [the Plan] through legislature—supporting readings and committee process, the standing committees were on par with what I have seen before in terms of how many people came out—what was different was the large number of citizens that were there." The classic-planning module was followed, but significant citizen involvement added non-linear elements to a traditional system.

Geyer and Rihani (2010) acknowledge limits of knowledge and importance of learning as key elements of conscious, complex systems. In the Lake Simcoe case, diverse perspectives helped to illuminate knowledge gaps and the importance of learning from each other. Participants acknowledged that the degree of predictability was limited. This thinking led to the Plan being created in an adaptive management framework.

The creation of scientific and advisory committees made of government and non-government stakeholders was considered novel and effective. One government actor said "The committees were quite important to the [creation] process, the science committee may have been more important in the legislation phase, and then may have shifted to the other committee while developing the Plan. Having an independent science committee to produce facts and state of the art information went a long way to backing up arguments of the need for a Plan. There is a credibility there that benefited from the process" (P1). These new approaches were producing results.

While it would be easy to focus on citizens' stories of inclusion, government actors were also positive about this multi-stakeholder environment. P7 stated: "it's rare to see so many resources be pooled together in one project in the government. I attribute that to the Premier and the drive for a delivery date that was ambitious…I've never seen such engagement of stakeholders, and involvement with the committees...that was used with quick turn around and fed into cabinet." Exploratory approaches that probed for results—rather than rigid processes—were implicit in such statements.

Resilience

The complexity-related concept of resilience has emerged in many fields. A resilient ecosystem "can withstand shocks and rebuild itself when necessary. Resilience in social systems has the added capacity of humans to anticipate and plan for the future" (Resilience Alliance). Resilient communities "draw mainly on local resources, knowledge, and expertise when faced with an incident or crisis. Resilience, in turn, contributes to a community's organizational effectiveness and its ability to drive positive social and economic change over time" (Fournier, 2012: i).

Social resilience was an important theme for over half the citizen and government participants in the original research (Gallagher, 2012). For example, some citizens formed a cohesive unit to anticipate what would be needed to avoid future ecological disasters.

Tangibility

The theme of tangibility was emphasized by five government and three non-governmental participants in the original research. Non-governmental participants needed tangible research to support their claims about declining water quality, and to gain inclusion into the creation phase of the Plan. The role of science was emphasized in interviews. When citizens brought forward tangible issues and scientific data, it was easier for government actors to make the case

Participant	G=Government C=Citizen	M or F	Types of Systems[*]	Key Starting Condition	Predictability[**]	Resilience	Tangibility
1	G	M	P B E	LSEMS	O	Y	Y
3	G	F	R P	LSEMS	O	N	Y
4	G	M	P E	Combination: LSEMS, science, citizen involvement, funding…	O	Y	Y
7	G	F	P R E A B	Combination: citizen involvement, high profile of lake, political and funding commitments…	U	Y	Y
8	G	F	P			N	N
10	G	F	P E R S	Political Priority	O	N	Y
2	C	F	P B	Public Concern	O	Y	Y
5	C	F	H S P	Public Concern	U	Y	Y
6	C	F	E P S	Environmental Indicators	O	Y	N
9	C	M	B H P S	Combination of many: citizen involvement, connection to the lake, concern for home and identity	U	Y	Y
11	C	M	B H P R	LSEMS	U	N	N
12	C	M	H A P R S	Public Concern	U	Y	N
13	C	M	H P S	LSEMS	U	N	N
14	C	M	B H P R S	Connection to Lake Simcoe	U	N	N

Table 1 *Themes emphasized by each participant (Note: The boundaries theme is not included because of the diversity of content).*

* A = Agricultural B = Biological or Ecological. E = Economic. H = Human, Spiritual, Artistic. P = Political and Governance. R = Recreation and Tourism S = Social and Cultural. T = Science and Technology.
** O = relatively orderly approach, albeit with surprises. U = relatively unordered approach: more creative and responsive.

for a new Act: government was able to respond more effectively to non-governmental voices.

Summary

Table 1 shows which participants emphasized which themes. A blank cell indicates no emphasis.

Findings show that groups inside and outside government chose to gather around an environmental attractor for which scientific data had been gathered. Government groups opened up to new ways of working; non-government groups invested time and energy learning about how to push boundaries, develop patience without losing energy, engage in political processes, take on projects and galvanize action. Through interactions across many types of boundaries, an intellectual estuary™ emerged. Coevolution enabled all groups to make progress with at least some of their concerns. This coevolution flowed into use of an adaptive management strategy for implementation of the plan, which was coherent with complexity thinking. Often "surprises" in government are to be avoided, but participants framed surprises in a positive light.

Discussion And Conclusions

Watersheds And Mega-Regions: Common Ground?

Use of a watershed boundary was a key innovation in this policy work. Watersheds include all rivers, streams and wetlands that drain into—or are connected to—a body of water. From an ecological perspective, this ensured that all systems associated with Lake Simcoe would be under regulation from the Plan. However, a watershed boundary was not an intuitive choice for many residents, nor was it a familiar landscape for policy work. Moreover, a watershed focus brings governance challenges because of multiple value sets, ministries, municipalities, and jurisdictions.

Similar challenges are faced in mega-regions where multiple municipalities are gradually connecting. They, too, share "interdependencies in their economies, infrastructure, natural resources, and the welfare of their citizens" and can be considered a "complex system without a public entity that focuses on the overall welfare" (Innes *et al.*, 2011: 1).

The critical resource of fresh water can draw people into dialogue about governance. Innes and her coauthors conducted their complexity and governance research around two water-planning projects in California. There have been other forums such as: "A Water Gathering: Collaborative Watershed Governance in [British Columbia] and Beyond." Also in British Columbia, the Cowichan Watershed Board is a "recent addition to this growing number of watershed governance organizations and agencies across the country" (Brandes & Brandes, 2012: 18).

This intersection of water policy, sustainability and complexity thinking forms an important edge, from which we can learn and adapt our planning and governance processes. Findings from the California and Lake Simcoe studies are similar. Innes *et al.* concluded that successes for complex governance in California were rooted in:

> *diverse, interdependent players; collaborative dialogue; joint knowledge development; creation of networks and social and political capital; and boundary spanning. They were largely self-organizing, building capacity*

and altering norms and practices to focus on questions beyond the parochial interests of players. They created new and often long-term working relationships and a collective ability to respond constructively to changes and stresses on the system. (p.55)

Was Complexity Embraced In Lake Simcoe Policy Work?

Geyer and Rihani contrast traditional orderly and emerging complexity public policy perspectives (2010: 33-34). As examples, an orderly perspective involves: "duplicating traditional scientific knowledge and methods is the primary justification of orderly public policy," whereas a complexity perspective includes: "A flexible mix of traditional scientific and more qualitative, interpretive policy methods is the most effective strategy." Strategic implications of the orderly perspective: "The creation of an improved and stable order" contrast with the complexity perspective: "The key isn't to find the final order and implement it, but encourage the actors in the policy area to adapt and adjust to the continual evolutionary changes in their areas."

Based on interview data, government actors were comfortable with ordered perspectives but moved towards complexity perspectives over time. Citizens were less bound by tradition, and brought complexity perspectives into the process. This pattern has been observed in many organizations and groups (MacGillivray, 2009) and can be framed using Midgley's theory of boundary critique.

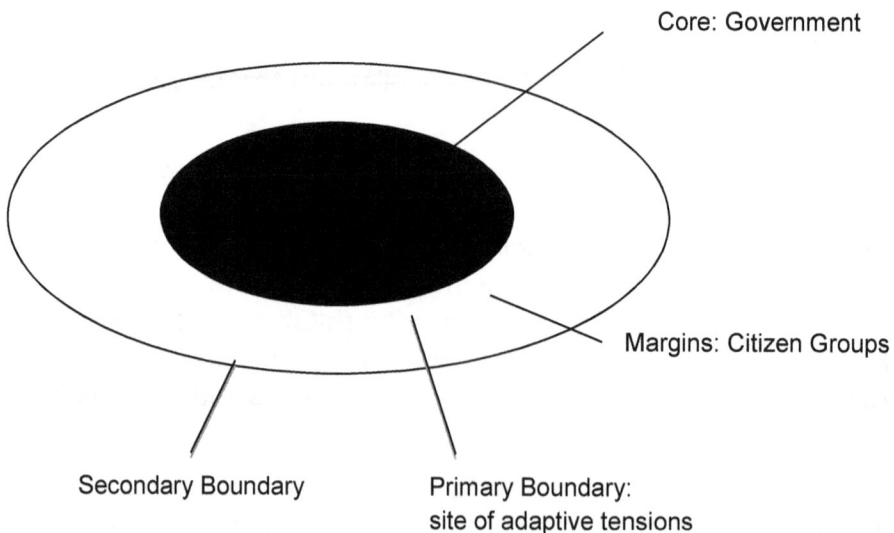

Figure 3 *Government as the Core, based on Midgley's Theory of Boundary Critique.*

To an extent, government actors were aware of the potential of complexity-based approaches. The eventual opening of the policy debate and inclusion of other viewpoints highlights the evolutionary change in the government sphere, and the gradual acceptance of new norms and methods of policy creation.

One strategy for boundary work between the core and margin is to generate adaptive tension (MacGillivray 2009: 181), which can "help people strive to problem-solve and improve" (189-190). Maguire and McKelvey (1999) describe adaptive tension as a gradient that can drive change from current to desired states. Tension therefore decreases with progress towards goals. Many participants described tensions around the primary boundary. Public servants were "getting a lot of pressure from various NGOs" (P1) and citizen groups (P2). Citizens were conscious of this: "Pressure from the public got the government to open up and move from the dark ages into opening the forum for public participation" (P5) and worked to raise awareness, media attention, energy and funds. These efforts were not only focused on legislation; citizens hoped for more complex outcomes including "stimulating innovation" (P2). Government realized over time that citizen involvement could enhance the quality of policy to the point where government aimed for a "gold standard" that could be "sold as a model" (P4).

Our interpretations of the data suggest that government did not approach this policy initiative from a complexity perspective, nor did the government participants proactively shift in that direction. But their approach did evolve. By way of contrast, many citizens had a complexity orientation from the start, as is typical with people operating in their private lives. They had to be tenacious, creative and resilient to be heard; using adaptive strategies that pushed the boundar-

	The Core (Government)	The Margins (Citizens)
Attractor	Scientific research was extensive, familiar, measurable, ordered and easily defensible	Scientific research might or might not best represent a citizen's views but it was a solid way into the process
Strategy	Somewhat ordered; inclusion of citizen groups introduced complexity	Much more complex. Included probes, experiments and adaptive responses
Enablers	Security of more direct communication with constituents. Precedent of [complex] adaptive management in ecological work	Diversity of groups. Less need to be concerned about public perceptions of experiments
Implementation	Passing of legislation could be interpreted as achievement of final order	Citizens worried that their status and inclusion might not continue through evolutionary changes
Boundary strategy	Government may be tempted to move on to next political/ economic win	Use of adaptive tension across primary boundary (MacGillivray, 2009) has been effective to date and may continue.

Table 2 *Patterns in Government and Citizen Policy Work.*

ies between their groups and government. The policy initiative began to manifest attributes of a complex system, including self-organization and emergence, leading to the creation of new order. Through this process citizens helped a new form of policy creation to emerge: government and citizens created new order together in a coevolutionary process. The perceived impacts of this evolution were very positive.

The application of inclusive, knowledge and communication-intensive complexity based approaches can enrich and even accelerate the policy creation process. Complexity approaches could potentially help overcome initial inertia and tensions often found in policy creation that includes multi-stakeholders and non-governmental participants.

Because government could default to more ordered policy perspectives, citizens were anxious to find ways to coevolve through implementation and monitoring processes. The emerging interest in governance across large geographic areas will provide opportunities to learn about possibilities for longer-term coevolution. Because some citizens are concerned that the emphasis for coevolution may wane within government, it will be important to give attention to ongoing learning. Given that the policy development process was exceedingly collaborative, the learning design could mimic that model. Initiatives such as Howard Rheingold's (2012) evolving media map for social media learning could provide inspiration for collaborative, customized mapping and implementation.

In summary, the Lake Simcoe case study illustrates the positive outcomes that are possible when government defaults to less rigid boundaries, and is open to working without pre-determined outcomes and checkpoints. Geyer and Rihani (2010) describe how a complexity perspective shifts traditional beliefs:

> In the public policy domain the dominant framework of the twentieth century was undoubtedly the traditional orderly perspective. With this perspective, states could show that they were acting in the best, most rational interests of their citizens and that citizens should do as they were told. Moreover, citizens could even convince themselves that they were being taken care of by experts who knew what to do. Hence, the populace could wash its hands of individual responsibility in the relaxing allure of belief in order. Complexity undermines both the dominance of elites and the passivity of local actors. (p. 186)

The Lake Simcoe Plan brought this concept to life, as described by P5: "In the past people thought something was wrong with the lake and the government should fix it. This added up to the lake being our responsibility."

Acknowledgements

We would like to acknowledge Royal Roads and Fielding Graduate Universities for supporting the research and thinking behind this paper, the participants, and all the scholars who continue to attend to both theoretical and practice elements of complexity theory.

References

Boulton, J. (2010). "Complexity theory and implications for policy development," *Emergence: Complexity & Organization*, ISSN 1532-7000, 12(2): 31-40.

Brandes, O. and Brandes, L. (2012). "Think like a watershed: How would watershed governance change if we made decisions from an ecosystem perspective?" Water Canada, ISSN 1715-670X, (May/June): 16-19.

Churchman, C. (1998). "Poverty and development," *Human Systems Management*, ISSN 0167-2533, 17(1): 9-13.

Conn, E. (2010). "Community engagement in the social eco-system dance: Tools for practitioners," Third Sector Research Centre Discussion Paper, http://www.tsrc.ac.uk/LinkClick.aspx?fileticket=K8%2brbdUTghQ%3d&tabid=827.

Dennard, L.F., Richardson, K.A. and Morçöl, G. (eds.) (2008). *Complexity and Policy Analysis: Tools and Concepts for Designing Robust Policies in a Complex World*, ISBN 9780981703220.

Fournier, S. (2012). "Getting it right: Assessing and building resilience in Canada's north," Conference Board of Canada report, 12-294.

Gallagher, K. (2012). "Environmental policy creation: Using the Lake Simcoe protection plan as a case study of success," Master's Thesis, http://dspace.royalroads.ca/docs/handle/10170/515.

Gerrits, L. (2010). "Public decision-making as coevolution," *Emergence: Complexity & Organization*, ISSN 1532-7000, 12(1): 19-28.

Geyer, R. and Rihani, S. (2010). *Complexity and Public Policy: A New Approach to 21st Century Politics, Policy and Society*, ISBN 9780415556637.

Head, B.W. (2008). "Three lenses of evidence-based policy," *The Australian Journal of Public Administration*, ISSN 1467-8500, 67(1): 1-11.

Innis, J.E., Booher, D.E. and Di Vittorio, S. (2010). "Strategies for mega-region governance: Collaborative dialogue, networks, and self-organization," *Journal of the American Planning Association*, ISSN 0194-4363, 77(1): 55-67.

MacGillivray, A. (2009). *Perceptions and Uses of Boundaries by Respected Leaders: A Trans-Disciplinary Inquiry*, PhD Dissertation, 3399314.

Maguire, S., and McKelvey, B. (1999). "Complexity and management: Moving from fad to firm foundations," *Emergence*, ISSN 1521-3250, 1(2): 19-62.

Midgley, G. (2000). *Systemic Intervention: Philosophy, Methodology and Practice*, ISBN 9780306464881.

Mitleton-Kelly, E. (2003). "Ten Principles of Complexity & Enabling Infrastructures," in E. Mitleton-Kelly (ed.), *Complex Systems and Evolutionary Perspectives of Organizations: The Application of Complexity Theory to Organizations*, ISBN 9780080439570.

Pawson, R., Greenhalgh, T., Harvey, G. and Walshe, K. (2005). "Realist review: A new method of systematic review designed for complex policy interventions," *Journal of Health Services Research and Policy*, ISSN 1758-1060, 10: 21-34.

Porter, T. and Derry, R. (2012). "Sustainability and business in a complex world," *Business & Society Review*, ISSN 0045-3609, 117(1): 33-53.

Resilience Alliance: "Research on resilience in social-ecological systems-A basis for sustainability" *Resilience*, http://www.resalliance.org/index.php/resilience.

Rheingold, H. (2012). "The media we use for co-learning," http://socialmediaclassroom.com/host/vircom/page/the-media-we-use-for-co-learning.

Ruddick, S. (1996). "Constructing difference in public spaces: Race, class, and gender as interlocking systems," *Urban Geography*, ISSN 0272-3638, 17(2): 132-151.

Ruhnaar, H., Dieperink, C. and Driessen, P. (2005). "Policy analysis for sustainable development: Complexities and methodological responses," paper for the *Workshop on Complexity and Policy Analysis*, Cork, Ireland, 22-24 June.

Warburton, R.N. and Warburton, W.P. (2004). " Canada needs better data for evidence-based policy: Inconsistencies between administrative and survey data on welfare dependence and education," Canadian Public Policy, ISSN 0317-0861, 30(3): 241-255.

Zussman, D. (2003). "Evidence-based policy making: Some observations of recent Canadian experience," Social Policy Journal of New Zealand, ISSN 1172-4382, (20): 65-71.

Alice MacGillivray holds degrees from Fielding Graduate University (PhD in Human and Organizational Systems; MA in Human Development) as well as an MA in Leadership and an undergraduate degree in natural sciences and communication from Canadian universities. She spent several years in the public sector, working with sustainability and education. In 2005, she left her position directing knowledge management programs at Royal Roads University to work as an independent consultant. Alice has presented at several international academic and professional conferences. Alice is interested in complexity thinking, especially as it relates to leadership and the intersections of natural and social systems.

Krista Gallagher holds a masters degree in Environmental Management from Royal Roads University, and is pursuing a career in water policy. Krista recently completed her MA thesis at Royal Roads University on the creation and implementation process of *The Lake Simcoe Protection Act*. She looks forward to continuing a professional and academic career in water policy and management. Krista is interested in exploring how social systems affect and enhance the political decision making process, and the impacts of meaningful non-governmental participation that is beginning to occur in policy formation. Krista holds a bachelor's degree in Environmental Studies and Political Science from the University of Toronto. She has published with the G8 Research Group at the University of Toronto, analyzing carbon capture and storage commitments of G8 nations.

Applied

Health Care Policy That Meets The Patient's Needs

Joachim Sturmberg
Departments of General Practice, Monash University, AUS
The Newcastle University, AUS

Healthcare policy in most countries is fragmented—the focus is on discrete diseases, on technical approaches and on specific domains. Many patients miss out on health care that addresses their specific circumstances, be it their medical, social or environmental needs. What is lacking is as much a coherent understanding of health and disease, as a policy framework that acknowledges the interconnected dimensions affecting the health of people, and that proposes strategies to facilitate the development of local solution to a meaningful global goal. This paper proposes the notion of a *health care vortex* as a pragmatic metaphor to shift policy *towards meeting the patient's needs.*

> *Don't always follow the crowd, because nobody goes there any more. It's too crowded.*

> Yogi Berra

Introduction

Healthcare policy in most countries is fragmented—the focus is on discrete diseases, on technical approaches and on specific domains. Many patients miss out on health care that addresses their specific circumstances, be it their medical, social or environmental needs. What is lacking is as much a coherent understanding of health and disease, as a policy framework that acknowledges the interconnected dimensions affecting the health of people.

These problems and their underlying complexities are becoming ever more obvious. Not surprisingly the call for this special edition aptly observes: "As the complexity of policy design and management has to match the complexity detected in the human activity systems, it is increasingly apparent that the current policy production process shall change; it has to align itself with the transformations occurring in society but also engage in leveraging those more socially viable."

This paper proposes a complexity model that highlights the interconnections between policy and practice. It argues that a functional integrated health system requires an agreed value-based focus or core value to drive the system. By adhering to the system's core value, agents can act in adaptive ways that allow best possible local solutions to emerge that best meet the patient's needs—a proposal that goes to where the crowds haven't been.

Basic Concepts

Fundamental to reforming healthcare are an understanding of some important underlying concepts—health and disease; patient needs; determinants of health and disease; and health care and disease management. In this paper these concepts and their complex relationships can only be described briefly, and interested readers are referred to the relevant literature.

Health and Disease

The definitions of health have always embraced a subjective and evaluative dimension (Sturmberg, 2012). Thus health, illness and dis-ease refer to three points on a subjective scale of experience, whereas disease defines the pathology and pathophysiology of organs, cells, proteins or genes (Lewis, 2003; Sturmberg, 2009; Sturmberg et al., 2010). Both can be better understood as resulting from perturbations of highly interconnected networks at multiple scales (del Sol et al., 2010). Epidemiology clearly shows that the experience of health, illness and disease follows a Pareto distribution—80% of people are in good enough health not to require any healthcare services, of the remaining 16% are helped by primary care services, 3.2% by secondary care services, leaving a mere 0.8% that need tertiary care. These distributions have been remarkably stable over time (White et al., 1961; Green et al., 2001; Sturmberg et al., 2012b) (see Figure 1).

Health is also a dynamic state, constantly adapting to the changing somatic, emotional, social and cognitive circumstances. These insights have led to the development of the somato-psycho-socio-semiotic[1] framework of health (Sturmberg, 2007; Sturmberg, 2009; Sturmberg et al., 2010). This model of health represents each of the four domains of health and some of its characteristic elements in each corner. These domains are representational and symbolic in nature and the model underscores the interconnected relationships impacting on a person's health experience (health representing a basin of attraction) (Figure 2—left). Over time these interactions will change, and when plotted track the many different health states experienced by a person over time, i.e., show the dynamic changes of the experiential health trajectory around its basin of attraction (Figure 2—right).

---

1. Semiosis refers to the interpretation of signs or signals. E.g., someone firmly pinching your arm results in a pressure receptor responding and sending an electrical impulse via a sensory nerve to the brain. The brain has to interpret this electric signal which then is interpreted to be a "painful stimulus".

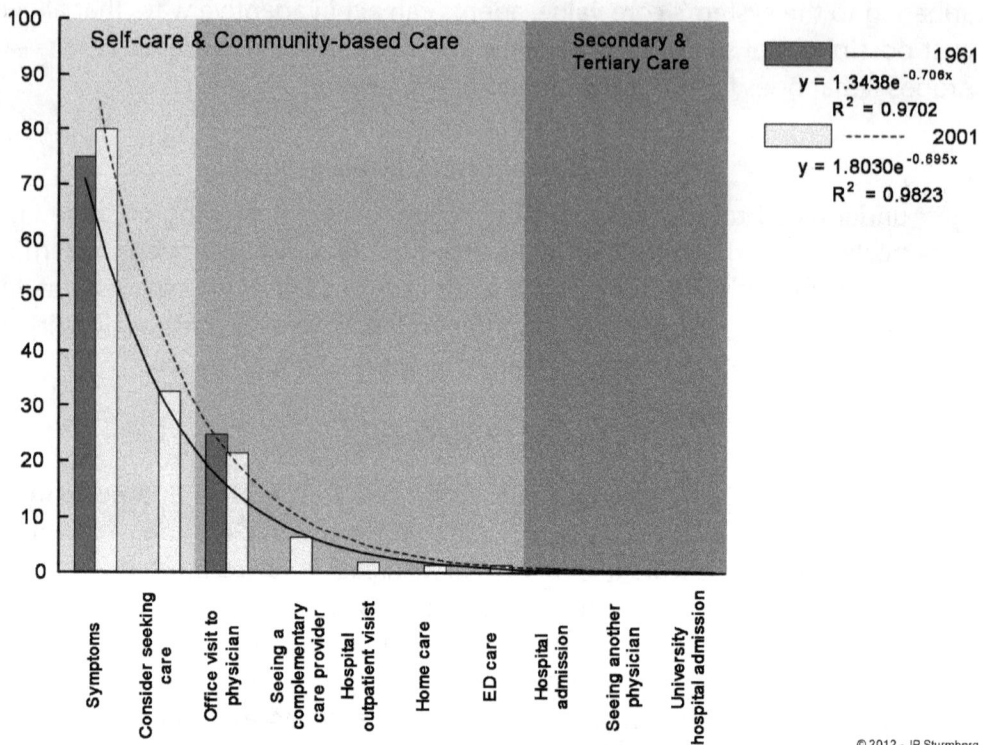

Figure 1 *Community Epidemiology of Health, Illness and Disease in 1961 and 2001.* The health, illness and disease experiences in the community follow the classic Pareto distribution (or 80:20 split).

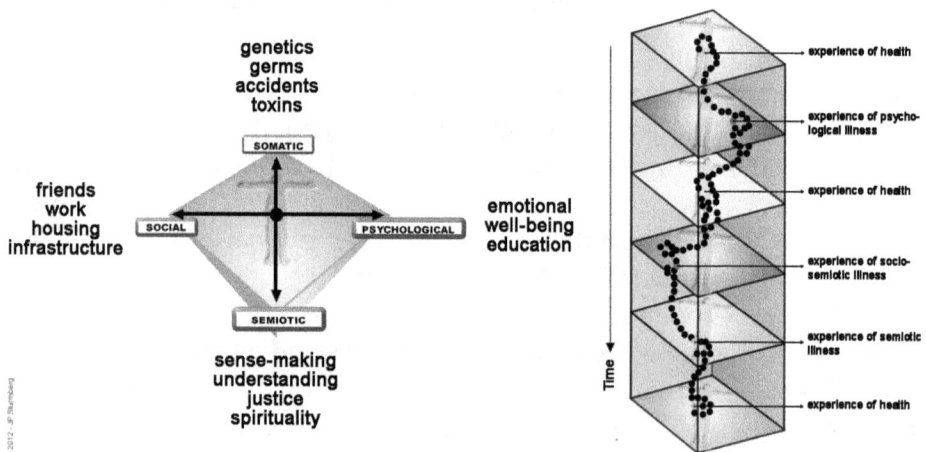

Figure 2 *The somato-psycho-socio-semiotic model of health and the health experience over time. The left panel shows the characteristic features of the four domains of the somato-psycho-socio-semiotic model of health. If the individual perceives the interactions between the elements in all domains being in balance, the person experiences health. The right panel shows the dynamic changes of a person's health experience over time which plots the trajectory of a person's basin of health attraction (the centre would represent a perfect health experience).*

E:CO Vol. 14 No. 4 2012 pp. 86-104

The importance of the health experience on morbidity and mortality has been described by Idler and Benyamini (Idler & Benyamini, 1997) and Jylhä (Jylhä, 2009); good self-rated health is positively correlated with significantly better health outcomes.

Patient Needs

Needs is probably the most controversial of these concepts. Maslow (1943) proposed that people have a hierarchy of psychological needs, ranging from basic physiological functions over security, love and belonging, and esteem to self-actualization, however, more recent research has shown that these hierarchies are more complicated and multi-faceted (Kenrick *et al.*, 2010). A political economy discussion of needs by Doyal and Gough (1984) emphasizes a much broader range of needs, including, besides of healthcare needs, adequate shelter, water and nutrition, as well as educational, environmental, social and relational domains.

Many of these needs domains are interrelated and interdependent, and related to the determinants of health. Appreciating the impact of *patient needs* is of significant clinical importance, as unmet patient needs stimulate the *physiological stress response* and contribute to poor health outcomes. If uninterrupted the circular relationship between unmet needs and poor health leads to a state of chronic poor health. The underlying physiological feedback loops and their relationship to health outcomes have been described widely in the psychoneuroimmunology literature (Kiecolt-Glaser *et al.*, 2002a,b; Ray, 2004; Bennett *et al.*, 2012).

In modern society needs are often confused with wants, perpetuated significantly by sensationalized and inaccurate publicity compounded by outright advertising by vested interest groups. The resulting exaggerated health anxieties lead to disease mongering (Heath, 2006; Moynihan & Henry, 2006; Herndon *et al.*, 2007; Moynihan *et al.*, 2008), over-diagnosis (Port *et al.*, 2000; Westin & Heath, 2005; Starfield *et al.*, 2008; Moynihan *et al.*, 2012) and associated inappropriate and/or unnecessary interventions.

Determinants Of Health And Disease

The WHO states:

> *Many factors combine together to affect the health of individuals and communities. Whether people are healthy or not, is determined by their circumstances and environment. To a large extent, factors such as where we live, the state of our environment, genetics, our income and education level, and our relationships with friends and family all have considerable impacts on health, whereas the more commonly considered factors such as access and use of health care services often have less of an impact.[2]*

Over and above the biological and health behavior aspects of the individual, the non-biological factors of that person's environment influence their health, illness and disease trajectories. Epidemiological studies have shown the significant impact of socio/economic/environmental factors on health and disease patterns (Kawachi *et al.*, 1997; Wallace & Wallace, 1997; Kawachi & Kennedy, 1999; Lynch *et al.*, 2001; Marmot, 2005, 2007). Not only do the impacts of these disadvantages persist in an individual moving up the social ladder (Thoits, 2010), they frequently are passed on via epigenetic mechanisms to future generations (Crews & McLachlan, 2006; Bird, 2007; Feinberg, 2007).

Health Care Or Disease Management

Health care and disease management are not necessarily opposites; under the right circumstances both can make valuable contributions to the patient's well-being. However there are important distinctions between the two, which have significant impacts on the process and outcome of healthcare. Principally health care focuses on care for the person's *experience of health* in light of their wishes and capabilities, whereas disease management focuses on ensuring that all providers and patients adhere to prescribed processes of management with the aim to achieve *predefined desirable measures of disease control[3]*.

Person-centred health care requires an understanding of the person and her illness, which requires time, or a certain slowness (Sturmberg & Cilliers, 2009), and an ongoing doctor-patient relationship (Fugelli, 1998; Sturmberg, 2007; Scott *et al.*, 2008). The highly dynamic and emergent processes of person-centred care ultimately achieve the economic prerogative of maximal effectiveness and optimal efficiency. In contrast disease management, being largely protocol driven, is linear and prescriptive in nature; it takes less time and can be delivered by anyone, increasing fragmentation of care. These characteristics, in the gaze of the prevailing industrial-economic paradigm, are, mistakenly, regarded as indicators of optimal efficiency but fail to take account of effectiveness.

---

2. http://www.who.int/hia/evidence/doh/en/.

3. Surrogate outcome measures are not the same as "real" outcome measures, and are open to misrepresentation/misinterpretation of the actual intervention findings/mechanisms.

E:CO Vol. 14 No. 4 2012 pp. 86-104

Health—The Central Focus Of The Healthcare System

How can we understand the many different domains that affect the health and health experience of patients and communities? Central to the healthcare system is its focus or goals, described as its core values, which should be the *health* of every patient. Core values are those that remain unchanged in a changing world, and drive all the structural requirements and interactions of the healthcare system. The metaphor of the health care vortex shall illustrate the structure and function of such a focused system.

The Vortex As A Metaphor To Understanding Complex Adaptive Systems

Capra (Capra, 1996) used the "bathtub" vortex to illustrate the principles of structure and function of complex adaptive systems. The structure of the vortex "spontaneously" emerges through self-organization when one opens the plughole. Even though a vortex exhibits highly complex dynamics, its structure remains remarkably stable.

The apex of a vortex provides the driving forces that maintain it. Interactions are greater and more complex closer to the apex and decrease towards its top. Disturbances of the vortex at the top cause limited disturbance of the whole structure, whereas disturbances closer to the apex result in major changes to the structure. As long as the central focus of the vortex is maintained, disturbances have only a temporary effect. The vortex's self-organizing properties will restore it closely back to its initial state.

The *Health Care Vortex*—A Metaphor to Understanding the Self-Organizing Dynamics of a *Complex Adaptive Health System*

Building on this metaphor we propose the *health care vortex* as the means to understanding the stability and dynamics of the health system. Like any other *complex adaptive system* the health system requires an attractor that "governs" its dynamic stability.

This attractor ought to be the *person's experience of good health*.

As shown above (refer to Figure 2) a person's health experience is the result of her unique internal complex adaptive dynamics. The person and the health system are interlinking systems; people's health experiences form the basin of attraction for a complex adaptive people-centred health care system.

As a corollary a complex adaptive people-centred health system needs to deliver interconnected *health maintaining* as well as *health restoring* structures and functions. These need to reflect both, the epidemiology of *good health* and the *need to seek care* irrespective of the presence or absence of identifiable biomedical changes, and the goal to return the patient to *her previous state of good personal health*.

Applying The Health Care Vortex Metaphor

Agents within large systems tend to self-organise according to functional domains within the system, which describe sub-systems. These sub-systems organise in hierarchies (Simon, 1962) which are interconnected and interdependent in line with the *system's attractor*. These functional domains can be described by the familiar hierarchical concepts of the macro (policy), meso (community), micro (practice) and nano (individual) levels.

Figure 3 attempts to portray the healthcare system in the vortex model. The vortex on the left describes the prevailing health system around an *industrial-economic* core driver, the vortex on the right the desirable health system around a *patient's health experience* core driver. The self-organizing dynamics of each complex adaptive system "exactly produce" the outcomes its driver "demands", both in terms of structure and function.

The Industrial-Economic Attractor

The industrial-economic attractor is based on a top-down command structure. Its basic belief is in breaking down every problem into small parts (fragmentation) as a means to most cost-effectively deal with it. It therefore is not unexpected that each level organizes and interacts in ways that deals with a small part (disease-specific protocols and guidelines), and promotes pay-for-performance strategies as a tool to achieve best possible financial efficiency, though profiteering or rationing emerge as unintended consequences.

This mode ignores the increasing complexity as one moves from the macro to the nano level within the vortex, and explains the "siloization" of issues and approaches. It also makes understandable the failure to realise the otherwise worthwhile goals of bettering the health system so it can achieve better health outcomes. In other words, this failure is neither unexpected nor unpredictable; it is the best that the system as a whole can achieve.

The Health Experience Attractor

Patient-centeredness and the *patient's health experience* have been at the forefront of medical care for millennia. They form the essence of good health care; however, this notion started to become undermined since Descartes' mind and body split in the 17th century. The scientific and technological discoveries of reductionist biomedical research over the past 50 years firmly shifted the focus away from the patient to "discrete diseases" (Pellegrino & Thomasma, 1981). Only in recent times have the need to focus on the patient's health experience and the salutogenic effects of patient-centeredness been "re-discovered".

Implementing a health experience attractor will allow structures and interactions to emerge that, under this given circumstance (initial condition), best meet the patient's needs and at the same time reflect overarching health policy goals.

E:CO Vol. 14 No. 4 2012 pp. 86-104

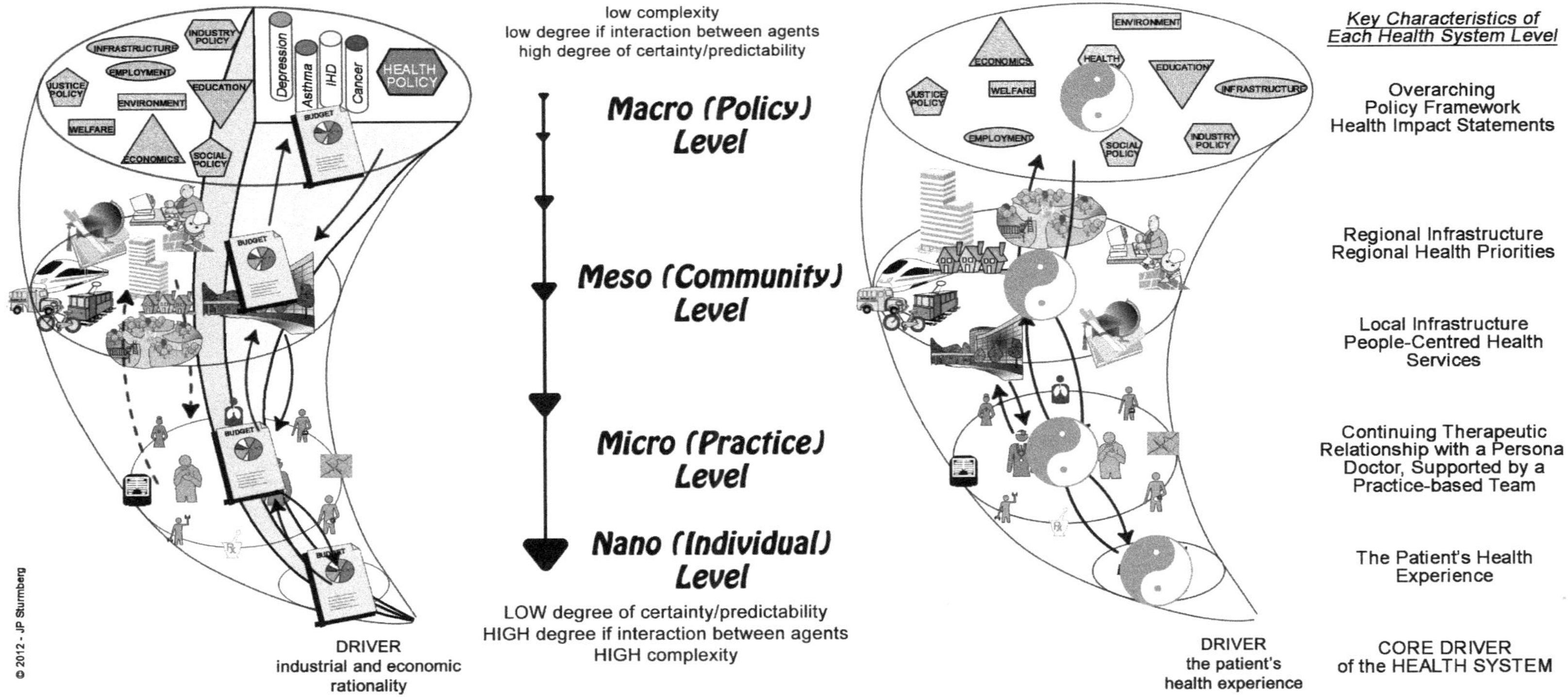

INDUSTRY POLICY
INFRASTRUCTURE
EMPLOYMENT
JUSTICE POLICY
ENVIRONMENT
EDUCATION
WELFARE
ECONOMICS
SOCIAL POLICY
Depression
Asthma
IHD
Cancer
HEALTH POLICY
BUDGET
© 2012 - JP Sturmberg
DRIVER
industrial and economic rationality
low complexity
low degree if interaction between agents
high degree of certainty/predictability
Macro (Policy) Level
Meso (Community) Level
Micro (Practice) Level
Nano (Individual) Level
LOW degree of certainty/predictability
HIGH degree if interaction between agents
HIGH complexity
ENVIRONMENT
HEALTH
DRIVER
the patient's
health experience
Key Characteristics of Each Health System Level
Overarching
Policy Framework
Health Impact Statements
Regional Infrastructure
Regional Health Priorities
Local Infrastructure
People-Centred Health Services
Continuing Therapeutic Relationship with a Persona Doctor, Supported by a Practice-based Team
The Patient's Health Experience
CORE DRIVER
of the HEALTH SYSTEM

Figure 3 *Summary of the structures and relationships of the current healthcare system (left) and the "ideal" personal health experience driven health care system (right). Different agents organise into sub-systems which are hierarchically ordered and commonly understood as policy, community, practice and individual levels of organization. The system dynamics increase from the top to the apex of the vortex, associated with an increase in complexity. The right hand column provides some examples of the characteristics and issues affecting each of the system levels.*

Complex adaptive systems will achieve multiple, mutually agreeable solutions, each being the best local system configuration to best support the efforts at the micro/nano (practice/individual) level. Its achievements need to feedback to the meso/macro (community/policy) level to allow for the "fine tuning" of system parameters.

Feedback between the system's agents, *all of whom are focused on the system's attractor*, allow the emergence of "the right action at the right point in time in this particular setting". Two examples shall illustrate the importance of the system's attractor for the functioning of the health system.

Unnecessary hospitalization and re-admission is a major economic and policy issue. An elderly patient living alone at home and suffering from several morbidities is at high risk of recurrent hospitalization. This patient will require a practice based team of health professionals who provide and monitor the management of the patient's specific conditions and coordinate other service providers, e.g. meals or community transport. However, this patient may equally need access to counselling by a psychologist, and community visitors to provide social support over the weekend when the patient is fearful of being "stuck alone at home" without ready access to any support (Figure 4—left panel) (Martin *et al.*, 2012).

Being truly patient-centred will also mean to properly assess and modify the impacts of socio/economic/environmental domains on the patient's health and illness experiences which currently are largely regarded as being outside the health profession's remit. The importance of these on the health of individuals and the community was already known in Roman times (don't build near a swamp, provide safe water, and each dwelling having a water closet), and the "Shape up Somerville" project[4] is a current example of a community wide participatory health improvement initiative in light of the growing obesity crisis in a diverse urban community (Figure 4—right panel) (Economos & Curtatone, 2010).

Policy Implications

To reiterate, a vortex organizes around its core driver or attractor—*a good personal health experience*. The importance of core values on the functions and outcomes of a system need to be underlined again. *Core values are those that remain unchanged in a changing world*: they do not change in response to market, financial or administrative changes[5], and sustain the organization in times of challenge. Values and ethics are the "soul" of what an organization stands for, how it conducts itself, and *how it guides the behavior of its members*.

---

4. The key features of the project are summarized on the Somerville website (http://www.somervillema.gov/departments/health/sus).

5. What are core values? http://www.nps.gov/training/uc/whcv.htm, How Will Core Values be Used? http://www.nps.gov/training/uc/hwcvbu.htm.

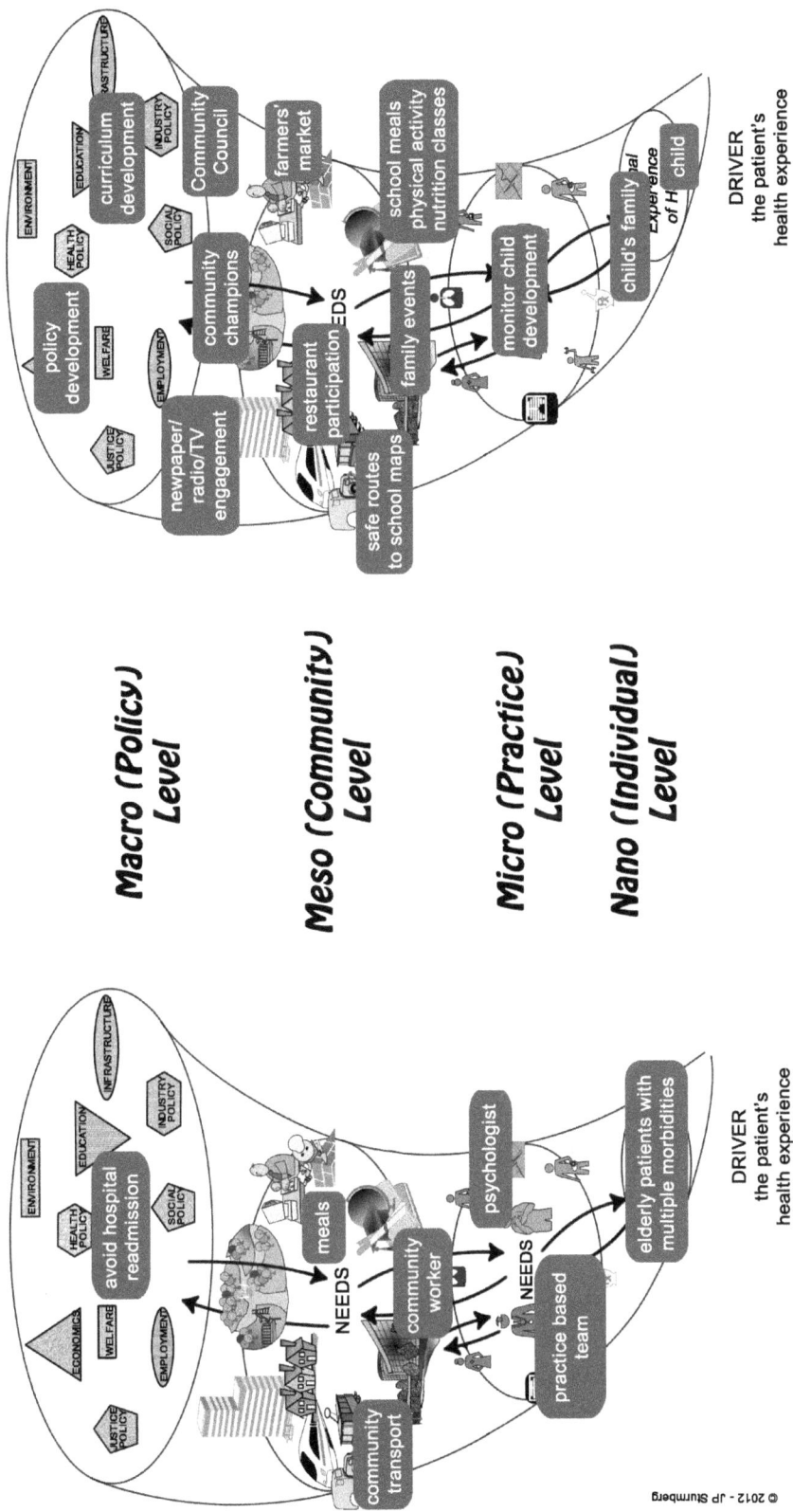

Figure 4 *The patient health experience attractor "in action" within the health care vortex. The left panel describes the interlinking agents "in action" to prevent unnecessary hospitalization of frail elderly patients in the community, based on Martin et als. work, the right panel describes the interlinking agents "in action" to prevent obesity in the Somerville community based on Economos et als. work. The latter in particular illustrates how the focus on the attractor results in "synchronous" actions and interactions between agents in a sub-system or level, and interactions across levels.*

Being the core driver of the entire system they guide the work an organization does, individually and collectively, and provide the moral compass that guides *how the necessary work gets done*. They define the "governing principles" of the organization and uniformly apply to all of its organizational units (Sturmberg *et al.*, 2012a).

To pre-empt any confusion it needs to be emphasized that core values are not descriptions of the work an organization does or the strategies it employs to accomplish its mission, i.e., they are not operating practices, business strategies, cultural norms, goals or competencies (Sturmberg *et al.*, 2012a).

As discussed above activities/changes in activities closest to the apex of the vortex are more influential on the functioning of the system than those closer to its top. Hence top-down policy changes, the prevailing way, will have limited impact on changes at the frontline service provision to people with highly variable illness and disease presentations, whereas allowing a bottom-up self-organizing approach at the service provision level will result in the most adapted service organization and will deliver more relevant improvements to the health needs and the health experiences of patients and communities.

Hence it is important to know at what part/level of the health care vortex changes are made, and *how* these changes interact with the other parts/levels. System dynamics (Sterman, 2000) taught us that many large changes/perturbations result in no change, whereas small changes/perturbations can result in major changes to the system. It all depends on making the "right" change at the "right" level and allowing the "temporal" dimension of change to transpire before intervening again (Dörner, 1996).

Outcomes are obviously an important part of policy evaluation. The overriding outcome, over and above any measurable change in discrete disease indicators, is the *patient's experience of health*. Interventions, however well intended, that decrease the patient's health experience are likely to be misplaced (Dooley, 1997).

Implications For The Macro Level

Policy makers have a remit to act in the interest of all people and communities. They need to appreciate that their leadership role is, to use Ron Heifetz's words (Heifetz, 1994), to facilitate the necessary adaptive work that has to be done by all at the grass roots level. This means that they have to support people at the grass roots level to work out their own solutions in line with the system's attractor, provide them with a sense of security and entitlement to make the necessary decisions that best fit their local environment, and that they are protected in case of having made a mistake. Mistakes are the only way to learn, and learning is essential for growing resilience and improving effectiveness in complex and uncertain environments.

E:CO Vol. 14 No. 4 2012 pp. 86-104

Pragmatically, health care reform primarily is core value (or focus) reform. It is the role of policy makers to facilitate and support the shift, as well as to accept that this shift will necessarily result in different but mutually agreeable solutions that reflect the best local outcome of the overarching goals.

Health policy makers also need to constantly collaborate with other policy areas, ensuring that these policy initiatives have fully considered all possible health consequences; clearly no policy should go ahead without a health impact statement.

Implications For The Meso Level

The meso level actors are charged to attend to community wide issues. These issues are broad, including structural domains like housing, education, work and social infrastructure, roads and open spaces, as well as relational interactions like health-related behaviors and social cohesiveness.

Agents at this level need to engage with their communities, they have to collaboratively approach issues with their community, and they need to think in 10-year time frames for any small change to result in a major outcome. System dynamics, beside of multidirectional feedback, have temporal characteristics that delay the emergence of important outcomes.

A longitudinal study demonstrates these system characteristics in the context of an underprivileged suburb (Sweet, 2011). People living in this suburb perceived the problems affecting them quite differently to those who look in from the outside; small things like vandalism and potholes affected their well-being more than the big-picture items of unemployment and housing. The development of skills and capacity of the community has been achieved over time, on the back of small projects that engaged people in their agendas, supported by engagement and cross-service collaboration. The conclusions provide a succinct summary about the characteristics of meso level thinking and approaches: [they must be] *willing to understand multiple perspectives, work with complexity in all its challenging messiness, and recognise the importance of both local detail and the over-arching big picture. It's no small ask in this age of the quick fix and the sound bite.*

Implications For The Micro Level

The micro level is the most dynamic, and the one with the greatest potential for improving or diminishing the *patient's health experience*. The level is least bounded, and its interconnections are truly complex; the actors bringing their own understandings of health, illness and disease, and moral norms and social expectations to the clinical encounter; the interactions of the therapeutic relationship trigger cognitive evaluative as well as unconscious responses fostering or inhibiting self-healing pathways; and all of these are interrelated to the contextual personal and social environment.

It is the level that experiences the greatest variability in patient need, in community capacity, and of staff capabilities. The high interdependence of these three domains has significant consequences for the structure and the relationships of a local health service (a practice/health centre/hospital). Being the delivery level of health services it is the level most scrutinised for its actions and outcomes, and it is the level most vulnerable to unjustly being "evaluated" as ineffective based on the typically used "measures of procedural, surrogate clinical and economic criteria".

Leadership is most widely distributed across the micro level, with all agents providing leadership in different ways at different times. As this level is typically organizationally structured, the main function of an organizational leader ought to be maintaining all agents' focus on the health care systems' core value, and this should be one of her main KPIs[6]. Individual service providers need to provide leadership in the consultation that achieves a joint understanding of the patient's needs and facilitates an improvement in health experience. The third aspect of collective leadership relates to understanding the community based health promoting/destroying factors and engaging with those agencies.

Outcomes at this level are least predictable, and indeed often surprisingly paradox. Responsive leaders should demand from their staff ongoing reflection on all outcomes to foster learning and adaptation in its constantly changing environment. Such adaptation, metaphorically described as "muddling through", is a complex management approach, which Lindblom defined as making decisions based on *successive limited comparisons,* whilst being aware that one only ever can achieve part of one's goals, and hopeful being able to avoid as many unanticipated consequences as possible (Lindblom, 1959); pragmatism matches legitimacy.

Implications For The Nano Level

The nano level is about the person, who at times will become a patient. The complex external and internal relationships and interactions of daily life shape a person's experience of health and illness. Taking care of oneself has important impacts on one's current and future experiences of health independent of the occurrence of—usually unavoidable—diseases, or increasing frailty. Psychoneuroimmunology has provided an understanding of the physiological mechanisms of care and self-care on health and health experience, as well as having provided an understanding of the disease-promoting mechanisms of undue life stresses on physical, emotional, social and cognitive well-being.

---

6. Key performance indicators.

It is essential to promote personal responsibility for health at this level, and "making it easy" to implement "common sense" lifestyle choices: healthy foods and physical activities to maintain an appropriate weight; abstaining from smoking and drug use; and moderation of alcohol consumption.

Of equal importance to good health are the establishment of strong social networks as they provide the important sense of belonging to one's neighborhood and communities. It is well established that social disintegration significantly threatens people's health and health experiences (Wallace & Wallace, 1997).

Conclusions

Each person is an intricate complex adaptive system who lives in a highly adaptive complex environment. Health and illness are continuous rather than discrete states that occur as much in the presence as absence of identifiable disease. Only those people who experience illness are going to have a need to engage with the health care system.

Health and illness are socially constructed, and so are society's responses in form of its healthcare system. This paper attempts to realign the notions of health, illness and disease with those of an effective and efficient health care system. This can be achieved by having the *patient's health experience and resulting needs* at the centre of the system, being the attractor that configures all of the health systems' agents and their relationships so they can work together to restore the patient's *experience of good health*.

The "top-down" policy layer, in the first instance, has to grabble with the need to shift the health system's attractor. Implementing this shift will mean to engage all agents in a process of rethinking the meaning of health and illness, the roles, goals and purpose of medical interventions on the illness trajectories, and the balance between patients' and health professionals' responsibilities in care. The longterm responsibility of policy makers is to firstly maintain all agents' focus on the system's attractor so that they can do the necessary adaptive work at the local level, and secondly to provide support to sustain the responsibilities and accountabilities across the system domains.

"Bottom up" strategies of local engagement and interprofessional discourse are required to *meeting the needs of the patient*. The adaptive work of health policy makers requires relinquishing power and control, the adaptive work for health service providers is to take greater responsibility of assessing local needs at the individual and community level, and to develop and implement solutions collaboratively with these affected agents.

The approaches and solutions described offer a "generic means" to understand the structure and function of complex adaptive systems. Of greatest importance is an understanding of the role of the attractor in "governing" the function of the system; a "wrong" attractor—very simply—never will achieve the "right" structures, dynamics or outcomes.

This special edition called for reflections on the need "*to align* [health policy] *with the transformations occurring in society*" and "[to leverage] *those* [that are] *more socially viable.*" Albert Schweitzer's insight succinctly summarise why we need to shift towards health policy that meets the needs of the patient and her health experience:

> *The witch doctor succeeds for the same reason that all of us succeed. Each patient carries his own doctor inside of him. They come to us not knowing that truth. We are at our best when we give the doctor who resides within each patient a chance to go to work* (Cousins, 1985).

References

Bennett, J.M., Gillie, B.L., Lindgren, M.E., Fagundes, C.P., and Kiecolt-Glaser, J.K. (2012). "Inflammation through a Psychoneuroimmunological Lens," in J.P. Sturmberg and C.M. Martin (eds.), *Handbook on Systems and Complexity in Health*, ISBN 9781461449973, pp. 279-300.

Bird, A. (2007). "Perceptions of epigenetics," *Nature*, ISSN 0028-0836, 447(7143): 396-398.

Capra, F. (1996). *The Web of Life*, ISBN 9780006547518.

Cousins, N. (1985). *Albert Schweitzer's Mission: Healing and Peace*, ISBN 9780393022384.

Crews, D. and McLachlan, J.A. (2006). "Epigenetics, evolution, endocrine disruption, health, and disease," *Endocrinology*, ISSN 0013-7227, 147(6): s4-10.

del Sol, A., Balling, R., Hood, L., and Galas, D. (2010). "Diseases as network perturbations," *Current Opinion in Biotechnology*, ISSN 0958-1669, 21(4): 566-571.

Dooley, K.J. (1997). "A complex adaptive systems model of organization change," *Nonlinear Dynamics, Psychology, and Life Sciences*, ISSN 1090-0578, 1(1): 69-97.

Dörner, D. (1996). *The Logic of Failure: Recognizing and Avoiding Error in Complex Situations*, ISBN 9780201479485.

Doyal, L. and Gough, I. (1984). "A theory of human needs," *Critical Social Policy*, ISSN 0261-0183, 4(10): 6-38.

Economos, C.D., and Curtatone, J.A. (2010). "Shaping up Somerville: A community initiative in Massachusetts," *Preventive Medicine*, ISSN 0091-7435, 50(Supplement): S97-S98.

Feinberg, A.P. (2007). "Phenotypic plasticity and the epigenetics of human disease," *Nature*, ISSN 0028-0836, 447(7143): 433-440.

Fugelli, P. (1998). "Clinical practice: between Aristotle and Cochrane," *Schweizerische Medizinische Wochenschrift - Supplementum*, ISSN 0250-5525, 128: 184-188.

Green, L., Fryer, G., Yawn, B., Lanier, D., and Dovey, S. (2001). "The ecology of medical care revisited," *New England Journal of Medicine*, ISSN 0028-4793, 344(26): 2021-2025.

Heath, I. (2006). "Combating Disease Mongering: Daunting but Nonetheless Essential," *PLoS Medicine*, ISSN 1549-1277, 3(4): e146.

Heifetz, R. (1994). *Leadership without Easy Answers*, ISBN 9780674518582.

Herndon, M.B., Schwartz, L.M., Woloshin, S., and Welch, H.G. (2007). "Implications of expanding disease definitions: The case of osteoporosis," *Health Affairs*, ISSN 0278-2715, 26(6): 1702-1711.

Idler, E.L., and Benyamini, Y. (1997). "Self-rated health and mortality: A review of twenty-seven community studies," *Journal of Health and Social Behavior*, ISSN 0022-1465, 38(1): 21-37.

Jylhä, M. (2009). "What is self-rated health and why does it predict mortality? Towards a unified conceptual model," *Social Science and Medicine*, ISSN 0277-9536, 69(3): 307-316.

Kawachi, I. and Kennedy, B. (1999). "Income inequality and health: Pathways and mechanisms," *Journal of Health Services Research and Policy*, ISSN 1355-8196, 34(1, Part II): 215-227.

Kawachi, I., Kennedy, B., Lochner, K., and Prothrow-Stith, D. (1997). "Social capital, income inequality, and mortality," *American Journal of Public Health*, ISSN 0090-0036, 87(9): 1491-1498.

Kenrick, D.T., Griskevicius, V., Neuberg, S.L., and Schaller, M. (2010). "Renovating the pyramid of needs," *Perspectives on Psychological Science*, ISSN 1745-6916, 5(3): 292-314.

Kiecolt-Glaser, J.K., McGuire, L., Robles, T.F., and Glaser, R. (2002a). "Psychoneuroimmunology and psychosomatic medicine: Back to the future," *Psychosomatic Medicine*, ISSN 0033-3174, 64(1): 15-28.

Kiecolt-Glaser, J.K., McGuire, L., Robles, T.F., and Glaser, R. (2002b). "Psychoneuroimmunology: Psychological influences on immune function and health," *Journal of Consulting and Clinical Psychology*, ISSN 0022-006X, 70(3):537-547.

Lewis, S. (2003). "Exploring the biological meaning of disease and health," http://sites.google.com/site/sjlewis55/presentations/vienna2003.

Lindblom, C.E. (1959). "The science of 'muddling through,'" *Public Administration Review*, ISSN 1540-6210, 19(2): 79-88.

Lynch, J., Davey Smith, G., Hillemeier, M., Shaw, M., Raghunathan, T., and Kaplan, G. (2001). "Income inequality, the psychosocial environment, and health: Comparisons of wealthy nations," *Lancet*, ISSN 0099-5355, 358: 194-200.

Marmot, M. (2005). "Social determinants of health inequities," *Lancet*, ISSN 0099-5355, 365(9464): 1099-1104.

Marmot, M. (2007). "Achieving health equity: From root causes to fair outcomes," *Lancet*, ISSN 0099-5355, 370(9593): 1153-1163.

Martin, C. M., Vogel, C., Grady, D., Zarabzadeh, A., Hederman, L., Kellett, J., Smith, K., and O'Shea, B. (2012). "Implementation of complex adaptive chronic care: The Patient Journey Record System (PaJR)," *Journal of Evaluation in Clinical Practice*, ISSN 1356-1294, 18(6): in press.

Maslow, A.H. (1943). "A theory of human motivation," *Psychological Review*, ISSN 0033-295X, 50(4): 370-396.

Moynihan, R. and Henry, D. (2006). "Disease mongering," *PLoS Medicine*, ISSN 1549-1277, 3(4): e191.

Moynihan, R., Doran, E., and Henry, D. (2008). "Disease mongering is now part of the global health debate," *PLoS Medicine*, ISSN 1549-1277, 5(5): e106.

Moynihan, R., Doust, J., and Henry, D. (2012). "Preventing overdiagnosis: How to stop harming the healthy," *British Medical Journal*, ISSN 0959-8138, 344:19-23.

Pellegrino, E. and Thomasma, D. (1981). *A Philosophical Basis of Medical Practice: Towards a Philosophy and Ethic of the Healing Professions*, ISBN 9780195027907.

Port, S., Demer, L., Jennrich, R., Walter, D., and Garfinkel, A. (2000). "Systolic blood pressure and mortality," *Lancet*, ISSN 0099-5355, 355(3): 175-180.

Ray, O. (2004). "The revolutionary health science of psychoendoneuroimmunology: a new paradigm for understanding health and treating illness," *Annals of the New York Academy of Sciences*, ISSN 0077-8923, 1032: 35-51.

Scott, J.G., Cohen, D., DiCicco-Bloom, B., Miller, W.L., Stange, K.C., and Crabtree, B.F. (2008). "Understanding healing relationships in primary care," *Annals of Family Medicine*, ISSN 1544-1709, 6(4): 315-322.

Simon, H.A. (1962). "The architecture of complexity," *Proceedings of the American Philosophical Society*, ISSN 0003-049X, 106(6): 467-482.

Starfield, B., Hyde, J., Gervas, J., and Heath, I. (2008). "The concept of prevention: a good idea gone astray?" *Journal of Epidemiology and Community Health*, ISSN 1470-2738, 62(7): 580-583.

Sterman, J. (2000). *Business Dynamics. Systems Thinking and Modeling for a Complex World*, ISBN 9780072311358.

Sturmberg, J.P. (2007). *The Foundations of Primary Care: Daring to be Different*, ISBN 9781846190810.

Sturmberg, J.P. (2009). "The personal nature of health," *Journal of Evaluation in Clinical Practice*, ISSN 1365-2753, 15(4): 766-769.

Sturmberg, J.P. (2012). "Health: A personal complex-adaptive state," in J.P. Sturmberg and C.M. Martin (eds.), *Handbook of Systems and Complexity in Health*, ISBN 9781461449973.

Sturmberg, J.P., and Cilliers, P. (2009). "Time and the consultation: An argument for a 'certain slowness'," *Journal of Evaluation in Clinical Practice*, ISSN 1365-2753, 15(5): 881-885.

Sturmberg, J.P., Martin, C.M., and Moes, M. (2010). "Health at the Centre of Health Systems Reform: How philosophy can inform policy," *Perspectives in Biology and Medicine*, ISSN 0031-5982, 53(3): 341-356.

Sturmberg, J.P., O'Halloran, D.M., and Martin, C.M. (2012a). "Health care reform: The need for a complex adaptive systems approach," in J.P. Sturmberg and C.M. Martin, (eds.), *Handbook of Systems and Complexity in Health*, ISBN 9781461449973.

Sturmberg, J.P., O'Halloran, D.M., and Martin, C.M. (2012b). "Understanding health system reform: A complex adaptive systems perspective," *Journal of Evaluation in Clinical Practice*, ISSN 1365-2753, 18(1): 202-208.

Sweet, M. (2011). "Understanding Miller," http://inside.org.au/understanding-miller/, published 28 March.

Thoits, P.A. (2010). "Stress and health," *Journal of Health and Social Behavior*, ISSN 0022-1465, 51(1 suppl): S41-S53.

Wallace, R. and Wallace, B. (1997). "Socioeconomic determinants of health: Community marginalization and the diffusion of disease and disorder in the United States," *British Medical Journal*, ISSN 0959-8138, 314: 1341-1345.

Westin, S. and Heath, I. (2005). "Thresholds for normal blood pressure and serum cholesterol," *British Medical Journal*, ISSN 0959-8138, 330(7506): 1461-1462.

White, K., Williams, F., and Greenberg, B. (1961). "The ecology of medical care," *New England Journal of Medicine*, ISSN 0028-4793, 265(18): 885-892.

Joachim Sturmberg is Adjunct A/Prof of General Practice in the Department of General Practice, Monash University, Melbourne, and Conjoint A/Prof of General Practice in the Department of General Practice, The Newcastle University, Newcastle, Australia. I am a graduate from Lübeck Medical School, Germany, where I also completed my PhD. Since 1989 I work in an urban group practice on the NSW Central Coast, with a particular interest in the ongoing patient-centred care of patients with chronic disease and the elderly. In 1994 I started to pursue systems and complexity research with an inquiry into the effects of continuity of care on the care processes and outcomes. Since then my research has expanded and includes the areas of understanding the complex notion of health, health care and healthcare reform, showing that health is an interconnected multi-dimensional construct encompassing somatic, psychological, social and semiotic or sense-making domains, that health care has to embrace the patient's understanding of her health as the basis for effective and efficient care, and that an effective and efficient healthcare system ought to put the patient at the centre. I am joint chief editor of the Forum on Systems and Complexity in Medicine and Healthcare in the Journal of Evaluation in Clinical Practice. Together with Carmel Martin and Jim Price I chair the Complexity SIG in WONCA.

Applied

A Tao Complexity Tool: Inducing A Paradigm Shift In Policy-Making

Caroline Fu[1] *& Richard Bergeon*[2]
1 Gonzaga University, USA
2 Bergeon, Fu and Associates, USA

This article introduces a Tao Complexity Tool for local and global public policy-making and analysis across a multiplicity of human affairs. This tool provides policy-makers with an understanding of the basic undergirding meta-dynamics in complex situations. Rather than dissecting a situation into discrete material components and examining each component separately, this tool helps policy-makers induce a paradigm envisioning the whole as an *energy-being*. The energy-being, an *abstraction*, is a system of activating forces in constant interwoven motion emerging from the past into the present, as a manifestation of observable patterns. Understanding the energy in an at-the-moment *being* contributes to *knowing* and informs policy-making of *becoming*. This article, briefly describing the tool, provides its philosophical foundation, language basics for communicating thoughts, and examples illustrating its use in policy-making practice.

Introduction

This paper presents a Tao Complexity Tool (Fu & Bergeon, 2012) for public policy-making. Briefly, this article describes the theoretical foundation and operating mechanics of the tool; suggests a way to invoke paradigm shift for communicating thoughts; and illustrates how to use the tool for analyzing policy.

Public policy-making has been wrestled with for centuries by societies in resolving public issues. Policy, whether it takes the form of a contract or a set of governing principles/rules, is most efficacious for human welfare when shaped by the perspicacity of collective purpose and discernment. Nonetheless, even a well-intended policy is shadowed by perpetual complexity issues that defy displacement imposed by ever-changing beliefs in deeply rooted diverse societies. With temporal actuality leading to multiple relativistic "sense perceptions" (Einstein, 1922/2005: 2) and fear of uncertainty, people tend to interpret a policy in the most opportune manner conceiving inelastic-assumptions and paradigmatic-expectations. "Western thought, the belief in certainty" (Prigogine, 1997: 4), and fragmentation of perceived reality (Bohm, 1980) engender anomalous "singularity" (Siu, 1974: 290); and precipitate "policy paradox" (Stone, 2012: 10) in conflicting opinions and perplexing contentions.

A new paradigmatic approach to policy-making that "subsumes" emergent complexity and "resonates" with the spontaneity of *Nature*[1] (Siu, 1978: 85) seems to be warranted. "The new paradigm that is now emerging," enfolds complexity in "the world as an integrated whole rather than a dissociated collection of parts" (Capra 1975/1991: 324), mirrors *Tao*[2] spontaneity acting "in harmony with nature[Nature]" (p. 117). To rethink policy-making requires "paradigm shifts" (Kuhn, 1970: viii; Vaill, 1989: 112); a shift from examining a situation as "a mere linear procession of discrete entities" to analyzing "emergence" of the spatio-temporal flow corroborated by "the laws of nature, the determinate enduring entities" (Whitehead, 1925/1953: 93). Inducing a paradigm-shift from "linear" to "nonlinear" (Richardson, 2008: 44-46), from compartmental to whole, necessitates tools to stimulate new perspectives across a multiplicity of human endeavors. One tool consideration is employing "the formulation of the laws of nature within the range of low [macroscopic] energies" to elucidate "human existence" (Prigogine, 1997: 6) as energy manifestations of the omnipresent forces in Nature (Fu, 2008) or Natureforces.

Theoretical Foundation Of The Tool

The Tao Complexity Tool (Fu & Bergeon, 2012) is based on Tao philosophy and Nature's complementarity attributes (Lao-Tze & Chuang-Tze[3], 500BCE/1993). The tool draws upon cosmology/complexity science and Einsteinian/Newtonian physics to explain the Natureforce energy-flows undergirding social power dynamics (Russell, 1938) and policies. This section de-

---

1. *Nature*, in this article denotes the "primordial nature" forces of "the Universe" (Whitehead, 1933/1961: 253) affecting human phenomena, "embracing in its complexity, the natural entities" (Whitehead, 1920/1964:13); rather than extended uses of the word *nature* as type, intrinsic characteristic, or temperament of thing.

2. *Tao* (道), explained by modern philosophers, "Tao has two meanings, namely, Ontology and Cosmology" (Fu, 1953: 6). "Tao in the Sense of Ontology" is the abstraction of Tai-Ji ,"Great Supreme [太極]" (p. 6), "Supreme Ultimate" (Capra, 1975/1991: 107), and "primal beginning" (R. Wilhelm & Baynes, 1950/1997: 298). Being in Tao "has its inner reality and its evidences. It is devoid of action and of form. It may be transmitted, but cannot be received. It may be obtained, but cannot be seen" (Fu, 1953: 6). "Tao in the Sense of Cosmology" signifies, "There is no base where Tao comes from; there is no hole into which Tao goes. There is actuality but no place; and there is length but no beginning and end" (p. 6); akin to electrons in quantum reality. Tao regards chaos: "*A violent order is disorder*" and "*A great disorder is an order. These two things are one*" (Wallace Stevens in Briggs & Peat, 1989: 81). Tao philosophy explains Nature's magnetic polarities affecting human endeavors, resonating with Whitehead's (1929/1967) notion of abstractions of "primordial nature" (p. 46).

3. Lao-Tze and Chuang-Tze are spellings for the philosophers, 老子 and 莊子 in this article.

E:CO Vol. 14 No. 4 2012 pp. 105-123

scribes how using Nature's attributes, rather than scientific formulas, as methods to explain the abstractions of an entity's at-the-moment being. Those attributes are compared to the works of theorists in the fields of environmental social policy and organizational learning policy.

Synchronizing With Nature

Synchronizing with Nature is the essence of Eastern Tao philosophy, which binds Nature's authentication to human experience to provide a method for verifying Western scientific disciplines. Although, both Eastern philosophy and Western disciplines are based on and aimed at learning from Nature, they bifurcate in practice. "According to Chinese tradition, nature is spontaneous harmony;" while Western tradition, "speaking about 'laws of nature' would thus subject nature to some external authority" (Prigogine, 1997: 12-13) beyond human amenability. Nature "means 'what is by itself'" and "is governed by simple, knowable laws" (p. 12). Perceiving Nature's laws/policies "underpinning the dynamics," the "ever-changing flux of [complexity] patterns," and *perpetual novelty* and emergence" (Holland, 1998: 4) is pivotal to a policy-making paradigm shift.

Leveraging Nature's Attributes As Methods

Whitehead (1929/1967) explained "primordial nature" (p. 46) using "cosmological scheme" and "philosophical tradition" (p. 54). Employing the methods of *verifying-by-experience* with original "*Natural Knowledge,*" rather than *abstractions*[4] of human-made scientific-formulas, Whitehead (1920/1964) verified Einstein's method of adapting "the principles of mathematical physics to the form of the relativity principle" (pp. vi-vii). Capturing in "philosophic abstractions," the "whole being of substance" and experiences is "necessary for the method of thought" to confirm "the metaphysical substratum of these factors in nature" (pp. 20-21). Various experiences are "from more concrete elements of nature, namely, from events" (p. 33). Thus, "human experience" derives "some fundamental fact of nature" (p. 125) from mathematical physics and complexity science to explain Nature; while Nature explains emergent being to augment *knowing* and inform *becoming*. "We are concerned only with Nature" that is "with the object of perceptual knowledge, and not with the synthesis of the knower with the known" (Whitehead, 1919/2008: vii). In our prevalent paradigms, we take for

---

4. *Abstractions* represent the correlations of "concrete processes, factual extractions, thoughts, preoccupation, and conceptualization— explainable by laws of energy in physics" (Fu, 2008: 20). The "notion of physical energy, which is at the base of physics, must then be conceived as an abstraction from the complex energy, emotional and purposeful" (Whitehead, 1933/1961: 186). Abstractions differ in meaning and intention from metaphors. Metaphors offer illusory, syllogistic comparisons without defined correlations with phenomena at hand.

granted the material provision of Nature, yet disregard its dynamic forces under-lying every moment in life and use as method for resolving public issues.

The At-The-Moment Being In Tao

The at-the-moment being in Tao connotes "nature is spontaneous harmony;" while the "concept of a passive nature subject to deterministic and time-revers-ible laws is quite specific to the Western world" (Prigogine, 1997: 12). In Tao, everything is nondeterministic, like orbiting electrons in "quantum reality" (Zo-har & Marshall, 1994: 47), "in a state of transformation"—"In each moment the future becomes present and the present, past" (Wilhelm & Wilhelm 1956/1995: 26). The energy-being is an indivisible energy, fluid, temporal, non-static, trans-forming, and flowing. While the Eastern concept of Nature differs from the West-ern, there are also similarities. The "Greek 'Taoist' was Heraclitus of Ephesus," who shared Lao-Tze's thought, "not only the emphasis on continuous change, which he [Heraclitus] expressed in his famous saying 'Everything flows,' but also the notion that all changes are cyclic" (Capra, 1975/1991: 116). Echoing Whitehead (1925/1953), the whole energy-being is captured in an abstraction from "actual-ity of the past" and "potentiality of the future" (: 151), "bearing an enduring pat-tern, constitutes its specious present" (p. 104).

> *Within this specious present the event realizes [sic] itself as a totality, and also in so doing realizes itself as grouping together a number of aspects of its own temporal parts. One and the same pattern is realized in the total event, and is exhibited by each of these various parts through an aspect of each part grasped into the togetherness of the total event.* (pp. 104-105)

The enduring pattern in the specious present encompasses the earlier life-his-tory of the same pattern, "exhibited by its aspects in this total event" (p. 105). Thus, memory of life-history shapes a dominant pattern endowed with "an el-ement of value in its own antecedent environment" (p. 105). "This concrete prehension[sic], from within, of the life-history of an enduring fact is analyz-able [sic] into two abstractions" (p. 105). One is the "enduring entity which has emerged as a real matter of fact" and the other is the "individualized [sic] em-bodiment of the underlying energy of realization [sic]" (p. 105). Tao philosophi-cal inquiries can invoke paradigmatic awareness; the specious present of our energy-being from past to present that edifies our at-the-moment knowing as we enter future moments.

The Ancient Tao Philosophy

The ancient Tao philosophy (Lao-Tze, 500BCE/1891) facilitates comprehension of Nature's wisdom and cosmological complementarity of polar yin-yang *energy-flow*[5] (Fu, 2008). P. Y. Fu (1953) noted, "Before heaven and earth were, Tao existed by itself from all time," "from metaphysics, epistemology, and the theory of human life and end with politics" or Nature's policy (p. 6). Tao was later used to represent the yin-yang polarity (☯), which was conceptualized by 3000BCE Chinese Sages from ontological traces of Nature's phenomena (Fu, 2008). Mimicking the day-night change experience, yin fades as it folds and flows into yang; and vice versa. Earth's rotation creates the cosmological effect of the day/night cycle and the basis for yearly seasonal cycles; generates gravitational forces; and induces Nature's polar magnetic field. The polar magnetism produces both attracting and repelling forces that engender energy-flow movement, which is also found in quantum electromagnetic fields and human interactions. Those energy-flow movements are captured as abstractions of Nature's attributes in events; such as transforming mass into energy occurring when mass travels at the speed of light (Einstein & Infeld, 1938/1966). Nature's being and existence, was grasped as the "Tao in the sense of ontology," rooted in itself (Chuang-tze in Fu, 1953: 6). Being, from the Tao perspective, is energy-flow in quantum existence; "appearance comes from non-appearance; Beginning comes out of non-beginning (in accordance with Cosmology)" (p. 16). Perceiving concrete events as abstractions of energy-flow activated by the Natureforces, rather than substance and material worth, requires a paradigm-shift; that shift would invoke rethinking policy-making.

The Natureforces

The Natureforces, explained by Tao philosophy, represent the abstraction of "the system" of five essential Nature's "activating forces" (Wilhelm & Wilhelm 1956/1995: 103), interacting within an entity. The abstractions correspond to the Natureforce attributes: the distinctiveness/sameness, interactive movements, interdependency, and embodiment of Tao polar yin-yang complementarity. The Natureforces connote movement, direction, and existence, manifested as energy-flow and transformation—the essence of change. The "tendency of consciousness" as whole-being comes from "dynamic conceptualizations [that] are

---

5. **Energy-flow** is an abstraction describing an entity's at-the-moment being. It denotes what Chinese call ch'i (氣) to represent energy in Nature and human affairs. Ch'i, "a state of energy" whole-being "consciousness" (Wilhelm & Wilhelm 1956/1995: 293), signifies movements, interacting flows, and intermingling fluidity attributes. There is ch'i in everything; similar to Einstein's averment that *everything is energy, everything*. Different kinds of ch'i are always interacting. Ch'i is the energy state that manifests the "being" or "presence" of an entity, a person, organization, enterprise, collective, or human affair (Fu, 2008: 26-27). Each entity's ch'i is part of the Ch'i of the Universe (大氣).

prevalent in China" (p. 293). Being, as a state of fluid energy-flow, is undividable. Dividing energy abstraction distorts the essence of the whole. The varying strength of each activating Natureforce influences the whole-being, which manifests as unique at-the-moment energy-flow pattern of the entity's state of wellness. The ancient glyphs of the Natureforces recorded in the oldest Chinese doctrine, *Book of History* (Wilhelm & Baynes, 1950/1997), translate as: "Water, Plant, Fire, Soil, and Metal" (Fu & Bergeon, 2012). The tool uses the Natureforce concept as a basis to apperceive complexity in the being, be it person, organization, or country. Natureforces are respected by modern scientists.

Niels Bohr's Parallel Theory

Niels Bohr's parallel theory noted, "For a parallel to the lesson of atomic theory," we turn to "those kinds of epistemological problems with which already thinkers like the Buddha and Lao Tzu have been confronted" (Bohr in Capra, 1975/1991: 18). Bohr continued, "when trying to harmonize our position as spectators and actors in the great drama of existence" (p. 18), the relationships between matters become explainable once the energy-flow is introduced as connector/binder. Capra echoed, "the two foundations of twentieth-century physics—quantum theory and relativity theory—both force us to see the world very much in the way a Hindu, Buddhist or Taoist sees it" (p. 18). So too, "in events which are in some sense the ultimate substance of nature" (Whitehead, 1920/1964: 19), do we find the tangible and observable being explained by energy-flow.

Whitehead's Insights Into Aristotelian Concept Of Nature's Elements

Whitehead's insights into Aristotelian concept of Nature's elements (noting that "ether has been invented by modern science as the substratum of the events which are spread through space and time", Whitehead, 1920/1964: 18) presented the ether as a "muddled notion confusing many relations" (p. 18). Those elements identified by Ionic Greeks—"earth, water, air, fire, and matter, and finally ether, are related in direct succession" to "postulated characters of ultimate substrata of nature" (p. 19). Whitehead (1920/1964) informed "the scientific doctrine of matter is really a hybrid through which philosophy passed on its way to the refined Aristotelian concept of substance" (pp. 19-20). "[T]ime and space should be attributes of the substance"; and "substance is as a substratum for attributes" (p. 21). Just so, "Chinese do not emphasize 'substance' as mass;" rather, substance has relations to space and time that flow as energy (Fu, 2008). All entities are collective attributes "conceived as a state of energy" being (Wilhelm & Wilhelm, 1956/1995: 293).

Science Recognizes That Light Has Dual—Particle And Wave—Properties

Western doctrine viewed events scientifically as substance. Eastern doctrine saw events philosophically, as energy-flow; two valid views of the same phenomena. The Ionian and Chinese philosophers both regarded what was observable in nature as guidance to life. Whitehead (1920/1964) theorized that the Greek "first philosophy illegitimately transformed" an abstraction "necessary for the method of thought, into the metaphysical substratum of these factors in nature which in various senses are assigned to entities as their attributes" (p. 20). The Chinese sages' thoughts aligned with Whitehead's. While the Ionians identified four substances and struggled with matter and ether, the Chinese identified five activating forces.

The five Natureforces are observable in the incessant interactions they have with one another in unfolding issues and enfolding policy. The Tao Natureforce concept as a tool could serve universally across cultural, spatiotemporal, and other divides, locally and globally for policy-making. We see recognition of the five natures forces in tools already employed in policy-making, summarized in Table 2 below. Comparing the Tao Natureforces and Whitehead/Aristotelian Nature's elements to Senge's learning organization disciplines and SACS environmental forces, we find parallels.

Senge's Learning Organization Disciplines

Senge's learning organization disciplines (Senge, 1990) are easily described as five learning forces. The set of five disciplines leads to being and living as a learning organization: team-learning level-sets each team-member's contribution, personal-mastery strives for actualization of personal contribution, shared-vision illuminates the collective vision, mental-models purposes awareness of team-member's thinking, systems-thinking binds learning wholistically to the collective-organizational-being. The fifth discipline, systems-thinking, explicates the implicate order of being; requires continuing participation of other four disciplines to comprehend the complex whole.

The SACS Environmental Forces

The SACS environmental forces, presented by Castellani and Hafferty (2009), also parallel the five Natureforces. The SACS (sociology and complexity science) theoretical environment is composed of: the level-setting of social complexity, the development of complex science (striving), the notable recent methodological innovation of autopoiesis and self-organization (illuminating), the historical emergence of complexity science (grounding), and the binding of systems perspective in sociology (pp. 105-138).

Both tools establish the substance and then establish relationships. While events transform continuously and enfolds/emerges dynamically from-moment-to-moment in a time-continuum, the snapshots of at-the-moment wellbeing that they employ do not reveal recognizable dynamic energy-flow patterns. The Tao Natureforce tool identifies the emerging complexity patterns allowing policy-makers to rectify/amend a policy as well as assess the true well-being that undergirds the implicate order of society.

Activating Natureforces	Water	Plant	Fire	Soil	Metal
Aspects of the forces relevant to nature's behavioral attributes	Water forms solid, liquid, or vapor; adapts to shape; seeks to level flowing to fill open spaces.	Plant endeavors to grow tall, wide, and deep, creating roots, branches, leaves, fruits, and/or seeds.	Fire appears spontaneously; spreads through; casts off embers; provides light, spark, and heat.	Soil provides grounding for its inhabitants; nurtures though ravaged by wind, sand, and water.	Metal, when melted, forms veins, nuggets, tools, machines, and alloys with other metals.
Relevancies to human affairs	People and collectives work to meet needs; reach consensus; shapes for growth; and adapt to conditions.	People and collectives seek to grow capability; add capacity; explore, discover, and conceive new ideas in response to challenges.	People and collectives seek, gain, expand, and absorb knowledge to enlighten; motivate to spark; compete, consume, concur to win and shine.	People and collectives support efforts based on ideal, purpose, value, and goals with sustaining resources and knowledge.	People and collectives unite and form cohesive, capable, strong workforces to confront and learn from adversity.

Table 1 *The five activating Natureforces—the movements, attributes, and relevancies (Table 13.1. in Fu & Bergeon, 2012).*

Senge probed, "Could [George Bernard] Shaw's 'being a force of nature' relate to Bohm's 'participation' in the 'unfolding' of the implicate order" (Senge in Senge *et al.*, 1994: 47)? In "Bohm's view, the implicate order is continually 'unfolding' into what we experience as the manifest world, 'the explicate order'" and "human beings participate in this 'unfoldment'" (p. 47).

Tool Mechanics And Language For Communicating Thoughts

The tool provides a mechanism for discerning emerged complexity patterns in public issues; assessing abstractions of the multiple forces activating undergirding policy dynamics; and gaining clarity of patterns exhibited as energy-flow (Fu, 2008) on policy-making fields. Using the tool, we hope to induce a paradigm shift perceiving complex phenomena as energy-being transformed: discrete matters/substances become indivisible energy wholes. As the tool delves into energy-being, the Being is "endowed with higher insight and more perfect intelligence," allowing one to act "according to his[/her/Nature's] own free will" (Einstein in Prigogine, 1997: 13). The energy-being, consisting of discernible complexity patterns, emerges from the past into the present being (Wilhelm & Wilhelm, 1995).

The Five Activating Natureforces

The five activating Natureforces, translated from the original ancient Chinese texts, and their behavioral attributes and relevancies to human affairs are described in Table 1 (Table 13.1 in Fu & Bergeon, 2012: 235). Table 2 lists, using contemporary nomenclature, the attributes of activating forces in various circumstances. The list starts with classical representations of: Chinese scholarly

Activating Natureforces	Water	Plant	Fire	Soil	Metal
Chinese classic scholarly interpretations	Wisdom	Virtue	Etiquette	Trust	Integrity
Aristotelian/ Ionian Nature's elements	Water	Matter	Fire	Earth	Ether
Eliot's poetic elucidation	Sense	Fancy	Spirit	Abstention	Appetency
Leadership action verbs	Balance	Strive	Spark	Ground	Bind
Organizational learning forces/ disciplines	Team learning	Personal mastery	Shared vision	Mental models	Systems thinking
SACS environmental forces	Level-setting	Development	Notability	Historical-emergence	Systemic-binding
The singularly essential art of management	Level-setting	Striving	Illuminating	Purposing	Binding
Policy-making action verbs	Level-setting	Striving	Illuminating	Purposing	Binding

Table 2 *The five activating Natureforces in various circumstances.*

Activating Natureforces	Water	Plant	Fire	Soil	Metal
Movement descriptions selected from various translations to represent activating Natureforces	Learn, placate, drown, flood, wear, bypass, flow, adapt, scour, change, balance, quiet, cleanse, still, nurture	Invent, grow, innovate, strive, absorb, expand, challenge, achieve, resist, bend, surround, surrender, yield	Energize, burn, warm, peak, enlighten, raise, spark, cheer, compel, force, lift, illuminate, yearn, destroy	Stabilize, endure, care, nourish, smother, ground, demand, alter, transform, transmute, provide, gravitate	Harvest, succeed, force, conduct, bind, rivet, cut, connect, enchain, consolidate, limit, determine, reform, solidify, enrich, contract, sink, melt

Table 3 *Suggested verbs for the activating Natureforces*
(Table 13.3. in Fu & Bergeon, 2012)

Energy-flows contributing to states of being	Descriptions and effects on one another
Constructive Energy-flow (first major energy-flow) —Grow, Generating	Energy-flow attributes: Nurture, Reveal, Illuminate, Expand, Energize • Water nourishes Plant; • Plant fuels Fire; • Fire fertilizes Soil; • Soil nurtures Metal; • Metal enriches Water.
Exhaustive Energy-flow (shadow of the Constructive Energy-flow) —Entropic, Consume	Energy-flow attributes: Neglect, Undo, Decay, Contract, Unbalance • Water erodes Metal; • Metal depletes Soil; • Soil diminishes Fire; • Fire burns Plant; • Plant absorbs Water.
Destructive Energy-flow (second major energy-flow) —Decline, Collapse	Energy-flow attributes: Discourage, Prevent, Separate, Inhibit, Dilute • Water extinguishes Fire; • Fire melts Metal; • Metal chops Plant; • Plant loosens Soil; • Soil muddies Water.
Regenerative Energy-flow (shadow of the Destructive Energy-flow) —Agitate, Stimulate	Energy-flow attributes: Enhance, Challenge, Provoke, Perturb, Enable • Water quenches Soil; • Soil nurtures Plant; • Plant grasps Metal; • Metal stirs Fire; • Fire boils Water.

Table 4 *Energy-flows of five Natureforces*
(modified from Table 13.2. in Fu & Bergeon, 2012)

descriptions, Whitehead's (1920/1964: 20) Aristotelian/Ionian elemental forces, and Eliot's (1943/1971: 18) poetic interpretations. The modern representations include: Fu & Bergeon's (2012) leadership action verbs; Senge's (1990) organizational learning forces; SACS (sociology and complexity science) environment forces. The policy-making forces selected for this article are from Siu's (1978) singularly essential art of management; and Dennard, Richardson and Morçöl's (2008) policy-making models and modeling. One can select other words to better represent forces in each specific entity and situation. Additional nomenclature choices for the five activating Natureforces are suggested in Table 3.

The Nature's Energy-Flows Constitute The Language

The Nature's energy-flows constitute the language for communicating thoughts. Table 4 below, translated from the original Chinese text, depicts how the five activating Natureforces affect each other (modified from Table 13.2 in Fu & Bergeon, 2012), manifested as energy-being of an entity. The manifestations are described as energy-flows, two major opposites, Constructive and Destructive, and their respective shadows, Exhaustive and Regenerative. The four energy-flows form a language to articulate the energy-state of an entity.

Augmenting Table 4 above, Figure 1 depicts graphically the interactions between five activating Natureforces and illustrates the intermingling "Nature dipolar" (Whitehead, 1929/1967: 511) relationships in the yin-yang Tao presence. The arrows denote how each Natureforce influences (and is affected by) the other four Natureforces. The arrows do not imply a temporal process; rather, all five Natureforces act simultaneously and do so indivisibly. There are four kinds of arrows on the map—curving, dotted curving, straight, and dotted straight—to represent their interactions. Table 4 describes the four energy-flows—two major and two corresponding shadow energy-flows.

The four energy-flows, constructive, exhaustive, destructive; and regenerative, constitute the language describing basic energy-flow meta-dynamics of the whole, emerging with varying intensity of each of Natureforces. As shown in Figure 1, each Natureforce has four out-going arrows affecting others and four, in-coming, being-affected by others. Depending on the situation, each can be contributing to an entity's wellbeing-whole in eight different ways. For example: adequate Water nourishes Plant; excessive, drowns; and lacking, withers.

> Water relates to plant—when Water is strong, it nourishes Plant, Constructive Energy-flow; however, when Plant is strong, it absorbs Water, Exhaustive Energy-flow, depletes energy.

> Water relates to Fire—when Water is strong, it extinguishes Fire, Destructive Energy-flow; yet, when Fire is strong, it boils Water, Regenerative Energy-flow, and creates powerful steam energy.

Water relates to Soil—when Water is strong, it quenches Soil, Regenerative Energy-flow; yet, when Soil is strong, it muddies Water, Destructive Energy-flow.

Water relates to Metal—when Water is strong, it erodes Metal, Exhaustive Energy-flow; however when Metal is strong, it enriches Water, Constructive Energy-flow.

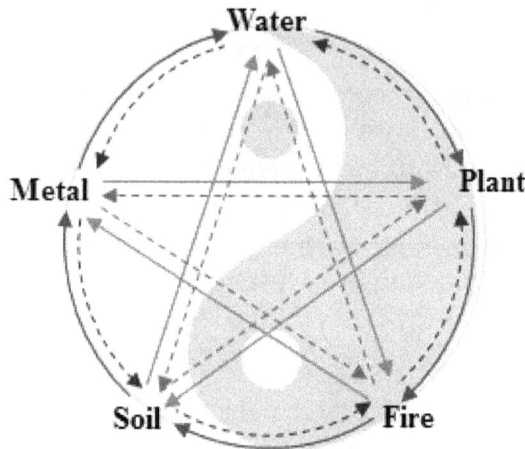

Figure 1 *Energy-flow diagram of the five activating Natureforces*

The Constructive Energy-flow (a major energy-flow) signifies that the entity is in a growth or generative state—a hopeful state without major obstacles. The curving solid arrows along the edge of the Tao symbol connect the five Natureforces clockwise:

Water > Plant > Fire > Soil > Metal > Water > . . .

The Exhaustive Energy-flow (the shadow opposite of the Constructive Energy-flow) denotes the entity is in a state of exhaustion or overconsumption—a warning about excessive attention to growth. The curving dotted arrows connect the five Natureforces counterclockwise:

Water > Metal > Soil > Fire > Plant > Water > . . .

The Destructive Energy-flow (the other major energy-flow) signifies that the entity is in decline or a collapsing state—an unfavorable situation; all Natureforces are destroying other Natureforces. The straight solid arrows, forming a solid-line star shape, inside the circles, connect the five Natureforces in a sequence (the end of an arrow is the beginning of the next arrow):

Water > Fire > Metal > Plant > Soil > Water > . . .

The Regenerative Energy-flow (the shadow opposite of the Destructive Energy-flow) denotes rising agitations in a situation, which compel interventions or rectifying actions. The straight dotted-line arrows alongside the straight solid-line arrows, but in the opposite direction of the solid lines, connect the five Natureforces forming a dotted-line star:

Water > Soil > Plant > Metal > Fire > Water > . . .

Thus, we experience the Nature's activating forces forming complex patterns from a mixture of the four partial or full energy-flows intertwined, moving and emerging. Partial energy-flows mean not all forces are fully engaged, or so weakened as to appear dormant or irrelevant. When we take a snapshot, we capture only the at-the-moment complexity pattern, appearing chaotic and unmanageable. Using the tool, we can discern those partial or full energy-flows involved in a complexity pattern. The five Natureforces involved in the constant performance of interactive functions as a system, manifested as the energy-being of an entity, be it individual, organization, or global enterprise,

Inducing Paradigm Shift In Policy-making

This tool facilitates paradigm shifts in policy-making from linear predictive thinking to nonlinear spontaneity. It employs a virtual playing field on which public policy issues and paradoxical concerns are played out and collective purpose, understood. "Public policies emerge from the nonlinear interactions of the human and natural realms and complexity theory can make a unique contribution to our understanding of such nonlinear and emergent phenomena" (Morçöl, 2008: 24). "Somehow, the simple laws of the agents generate an emergent behavior far beyond their individual capacities" and operate "without direction by a central executive," similar to electrons and neurons (Holland, 1998: 5).

The playing field provides containment for understanding public policy-making as abstractions of energy-flow manifestation of five activating forces constantly interacting with, and affecting, one another, creating an energy-flow of presence described as the at-the-moment being. The playing field mimics a "virtual presence is something which is not real in the space-time sense, yet it exerts a practical effect as if it were," like, "the square root of minus one" in mathematics (Siu, 1978: 88). "Yet this purely imaginary number is used very effectively in calculations involving real events;" without it, there would not "have been any modern physics;" or "the virtual presence" invented (p. 88). Being is never static, but emerging and transforming, continues to form collective meta-dynamics of energy-flow. Trying different scenarios by increasing/decreasing the intensity of one or more activating forces, one observes how the whole-being enfolds on the playing field.

As introduced earlier, the nomenclatures selected to represent the activating Natureforces for policy-making playing field are, "Level-setting," "Striving," "Illuminating," "Purposing," and "Binding" (Table 2). Level-setting has the water-leveling effect to learn and understand all at-that-moment public issues and concerns on the playing field. Striving is about considering all issues and reaching for the best achievable collective goals. Illuminating reveals the picture of the state of attainment. Purposing is about being purposeful with the collective wellness in mind. Binding is to wholistically consider and bind all aspects into a collective-policy-making-being.

The five activating forces are working in concert. Level-setting readies the stage for striving to achieve goals of understanding the issues. Striving for better understanding the goals illuminates the collective vision of wellness. Illuminating with vision of the collective wellness prepares policy-makers to be mindful about the purpose of the policy. Purposing about collective wellness induces new focus to bind all necessities. Binding all aspects together readies the playing field for attaining a higher-level understanding about public issues. That is an ideal Constructive Energy-flow at work shifting a policy-making playing field upward to facilitate attainment of best policy. However, given human fragmented reality views, social power grabbing, fear of uncertainty, and other factors would derail an entity from upward movement. The tool can be used to assess the energy-being on the playing field to understand which activating force was the culprit for derailing; then suggest intervening forces to get back on upward track.

Policy-making Examples

The following illustrate the tool usage for two textbook examples. Both show the ideal full Constructive Energy-flow, i.e., without a derailing the energy-flow into other partial Exhaustive, Destructive, or Regenerative patterns.

Policy-making: Science, Theory, Models, And Modeling

Dennard, Richardson and Morçöl (2008) introduced the complexity science of policy-making (pp. 1-22). We mapped the authors' points into the five activating forces interacting in the Constructive Energy-flow. This illustrates how the forces function in ideal policy-making exercises.

Level-setting prepares the social landscape to find the role of complexity science and research culture contributing to policy analysis (pp. 1-2), readying the social landscape for striving.

Striving enables attainment of a new social order and ethics to answer the postmodern critiques of policy analysis (pp. 3-7). Attaining new social orders leads to new policy enlightening.

Illuminating clarifies a new vision on policy analysis by theorizing, solving complex theoretical issues (pp. 10-12). The comprehension of theories roots purposeful policy.

Purposing translates rational planning to postmodern thinking and facilitates accountability exercises for both politicians and citizens (pp. 7-9, 13). Understanding the purpose of specific policies enables binding ethical accountability on globalized social landscapes.

E:CO Vol. 14 No. 4 2012 pp. 105-123

Binding provides modeling capability, a virtual playing field, for all stakeholder-citizens to develop robust policies (pp. 14-16). Modeling facilitates all activating forces to simulate different scenarios and enables level-setting of emerging horizontal landscapes.

Level-setting readjusts the old setting toward the emerging globalized landscape.

The Constructive Energy-flow repeats moving the work upwards until the collective policy purpose is reached. Each activating forces hold Tao yin/yang meta-dynamics in consciousness, maintaining the collective wellbeing.

Tao: The Singularly Essential Art Of Managing Policy-making Process

Siu's (1978) inconceivable Chinese Baseball game rules illustrate the uncertainty in life. In Chinese Baseball, "There is one and only one difference [from popular baseball-game]—after the ball leaves the pitcher's hand and as long as the ball is in the air, anyone can move any of the bases anywhere" (p. 84). It seems an impossible game to play; however, it lucidly describes the unpredictability and complexity we encounter daily. Adapting Siu's "Five Management Principles" (p. 84) as "The Singularly Essential Art" of policy-making involves "BIG decisions" (p. 83):

Level-setting: "*Act from an instantaneous apprehension of the totality*" (p. 85). Siu advised to think globally, employing a "wholist" strategy, act locally using "partist" strategy (p. 85), yet, consciously apprehend the totality of public issues. Level-setting between wholist and partist strategies prepares one to strive for the best attainable policy.

Striving: "*Subsume yourself and resonate*" (p. 86), Siu practiced this principle himself. "By being sensitive to the feelings of the people in the subsuming context" and "continually resonating" brings about "a shared mutuality" (p. 86). Thus, by attending to subsuming public issues and endeavoring to know people's needs, policy-making finds the resonance necessary to achieve its goals. Subsuming and resonating in achieving goals illuminates policy potentiality.

Illuminating: "*Maintain multiple tactical targets within attainable reach until the moment of final commitment*" (p. 86). Even those policies often considered as the best achievable may not be the most feasible until illumination brings about closure. Being enlightened by multiple policy choices and timing enables one to act purposefully.

Purposing: "*Be propitious*" (p. 87) and allow ample time to construct quality policy goals. People often seek immediate policy concurrence/resolution that result in a lifelong staccato of crisis-after-crisis. Those policies fail to lay the basis for the resolution of conflicts generating arguments for the sole purpose of winning. Acting elegantly enables policy-making to bind all to its collective purpose.

Binding: *"Orchestrate the virtual presences"* (p. 88). Create a policy playing field, where all stakeholders gather to analyze scenarios, gain deeper understanding of issues, and make policy affecting the collective. A "virtual presence in mathematics is the square root of minus one" (p. 88), producing practical answers that cannot be obtained in any other way. Orchestrating a virtual playing field enabled policy-making to handle the next level of complexity.

Level-setting: *"Act from an instantaneous apprehension of the totality"* (p. 85) to ready entities for obtaining best policies. The Constructive Energy-flow goes on until the policy goals are actualized.

Both examples given above are ideal the Constructive Energy-flow. In the real world, we often find minimal or excessive emphasis on one activating force that eventually causes the whole system to spin off into an unfavorable energy-flow. However, for each unfavorable energy-flow, we can find its Tao shadow opposite then, alleviate, readapt, and resume the intended policy-making dynamics.

Conclusion

Subsuming in the domain of public issues is an art; while finding resonance necessary to achieve policy-making goals is a science. The Tao Complexity Tool integrates science and art of policy-making to envision human affairs as energy-flow and transformation, rather than discrete instances of matter/mass. The tool helps induce paradigm shifts to rethink policy-making; consider the nonlinear dynamics of the activating forces at work within a collective whole instead of a linear procession of phenomena. On a virtual playing field, capturing real-world situation as abstractions of the imaginary specious present allows policy-makers to better understand the complex world. Being in harmony with Nature's meta-dynamics enables sensible policy-making and leads to the achievement of the stakeholders' collective purpose.

Practice Exercise

Envision a scenario that a policy was implemented without or with a weak collective purpose. Please use Figure 1 and Table 1, 2, and 3 as guides to consider: 1) Which Natureforce could be missing? 2) What could be the outcome? 3) What interventions could restore the policy-making effort to the original intention?

References

Bohm, D. (1980). *Wholeness and the Implicate Order*, ISBN 9780415289795 (2002).

Briggs, J. and Peat, E.D. (1989). *Turbulent Mirror: An Illustrated Guide to Chaos Theory and the Science of Wholeness*, ISBN 9780060160616.

Capra, F. (1975/1991). *The Tao of Physics: An Exploration of the Parallels Between Modern Physics and Eastern Mysticism*, ISBN 9780877735946.

Castellani, B. and Hafferty, F. W. (2009). *Sociology and Complexity Science: A New Field of Inquiry*, ISBN 9783540884613.

Dennard, L.F., Richardson, K.A., and Morçöl, G. (2008). "Introduction: Science, theory, models, and modeling," in L.F. Dennard, K.A. Richardson & G. Morçöl (eds.), *Complexity and Policy Analysis: Tools and Concepts for Designing Robust Policies in a Complex World*, ISBN 9780981703220. pp. 1-22.

Einstein, A. (2005). *The Meaning of Relativity, Including the Relativistic Theory of the Non-Symmetric Field*, ISBN 9781567311365 (1922).

Eliot, T.S. (1971). *Four Quartets*, ISBN 9780156332255.

Fu, C. (2008). *Energy-Flow: A New Perspective on James MacGregor Burns' Transforming Leadership: A New Pursuit of Happiness*, doctoral dissertation, Antioch University, http://rave.ohiolink.edu/etdc/view?acc_num=antioch1218205866.

Fu, C. and Bergeon, R. (2012). "A Tao Complexity Tool: Leading from Being," in J. Barbour (ed.), *Leading in Complex Worlds*, ISBN 9781118266991, pp. 227-251.

Fu, P.Y. (1953). *Philosophy of Chuangtse*, Master's thesis, National Taiwan University, Taipei.

Holland, J.H. (1998). *Emergence: From Chaos to Order*, ISBN 9780201149432.

Kuhn, T.S. (1970). *The Structure of Scientific Revolutions*, ISBN 9780226458045.

Lao-Tze and Chuang-Tze (1993). *The Essential Tao: An Initiation into the Heart of Taoism through the Authentic Tao Te Ching and the Inner Teachings of Chuang Tzu*, T. E. Cleary (trans.), original work published 500 BCE, ISBN 9780062502162.

Morçöl, G. (2008). "A complexity theory for Policy Analysis: An outline and proposals," in L.F. Dennard, K.A. Richardson and G. Morçöl (eds.), *Complexity and Policy Analysis: Tools and Concepts for Designing Robust Policies in a Complex World*, ISBN 9780981703220, pp. 23-36.

Prigogine, I. (1997). *The End of Certainty, Time, Chaos, and the New Laws of Nature*, ISBN 9780684837055.

Richardson, K.A. (2008). "On the limits of bottom-up computer simulation: Towards a nonlinear modeling culture," in L.F. Dennard, K.A. Richardson and G. Morçöl (eds.), *Complexity and Policy Analysis: Tools and Concepts for Designing Robust Policies in a Complex World*, ISBN 9780981703220, pp. 37-54.

Russell, B. (1938). *Power: A New Social Analysis*, ISBN 9780415094566.

Senge, P.M. (1990). *The Fifth Discipline, the Art & Practice of the Learning Organization*, ISBN 9780385260947.

Senge, P.M., Kleiner, A., Roberts, C., Ross, R.B., and Smith, B.J. (1994). *The Fifth Discipline Fieldbook: Strategies and Tools for Building a Learning Organization*, ISBN 9780385472562.

Siu, R.H.G. (1974). *Ch'i: A Neo-Taoist Approach to Life*, ISBN 9780262191234.

Siu, R.H.G. (1978). "SMR Forum: Management and the art of Chinese baseball," *Sloan Management Review*, ISSN 0019-848X, 19(3): 83-89.

Stone, D.A. (2012). *Policy Paradox: The Art of Political Decision Making*, ISBN 9780393912722.

Vaill, P.B. (1989). *Managing as a Performing Art, New Ideas for a World of Chaotic Change*, ISBN 9781555423698.

Whitehead, A.N. (1953). *Science and the Modern World*, originally-published 1925, ISBN 9780521237789 (2011).

Whitehead, A.N. (1961). *Adventures of Ideas*, originally-published 1933, ISBN 9780029351703 (1967).

Whitehead, A.N. (1964). *The Concept of Nature: The Tarner Lectures Delivered in Trinity College, November 1919*, originally-published 1920, ISBN 9781477660645 (2012).

Whitehead, A.N. (1967). *Process and Reality: An Essay in Cosmology*, originally-published 1929, ISBN 9780029345702 (1979).

Whitehead, A.N. (2008). *An Inquiry Concerning the Principles of Natural Knowledge*, originally-published 1919, ISBN 9781436555555.

Wilhelm, H. and Wilhelm, R. (1995). *Understanding the I Ching: The Wilhelm Lectures on the Book of Changes*, original work published 1956.

Wilhelm, R. and Baynes, C.F. (1997). *The I Ching or Book of Changes*, ISBN 9780691097503.

Zohar, D. and Marshall, I. (1994). *The Quantum Society: Mind, Physics, and a New Social Vision*, ISBN 9780688142308.

Caroline Fu is assistant professor of the Doctoral Program in Leadership Studies at Gonzaga University. She holds a PhD in Leadership and Change and an MA in Whole System Design, both from Antioch University. Her MS in Computer Sciences and BS in Applied Mathematics, Engineering (EE) and Physics are both from University of Wisconsin. She has a Certificate of Completion in System Dynamics Advanced Study from the Sloan School of Management, Massachusetts Institute of Technology. Her passion is furthering *Leadership as Energy-Flow* concepts for assessing complexity from a Tao philosophy and modern physics lens for leadership decision learning and global policy support.

Richard Bergeon is a principal of Bergeon, Fu and Associates. He holds a BBS in General Management (Wayne State University), an MA in Whole Systems Design (Antioch University), and certification as an Organization Systems Renewal Consultant. He has been employed as manager, director, and executive leadership consultant in banking, public utilities, transportation, manufacturing, pharmaceutics, and telecommunications, specializing in the adoption of new technologies and staff development. He is presently pursuing his doctoral degree at Gonzaga University, focusing on the role of values in global and intercultural leadership.

Philosophy

Complexity And Philosophy

Complexity, Acceleration, And Anticipation

Roberto Poli
Social Foresight and Department of Sociology and Social Research, University of Trento, ITA

Anticipatory governance is a system of prescriptions explicitly addressing the interplay among complexity, acceleration, and policy. Specifically, anticipatory governance provides a way to use foresight, networks, feedback and hierarchical links for the purpose of reducing risk and increasing the capacity to respond to events at their initial stages of development. In order to deal with acceleration, organizations must acquire a much greater sensitivity to weak signals concerning alternative futures and learn to respond them with increased flexibility and speed. The idea of anticipatory governance is expounded against a network of concepts and tools, including the difference between strong and weak signals, anticipatory systems, regulation, resilience, and the Foresight Maturity Model.

Seeds for a New Science

Willy-nilly, most decision-makers are positivists, and they regularly ask their consultants to give them 'solutions' able to solve problems once and for all. Complexity and the nature of contemporary science show that the claim to 'solve' (complex) problems is often ungrounded.

By way of an over-compressed summary, Newtonian science teaches us that natural systems are:

- Closed (only efficient causality is accepted; bottom-up, top-down, 'final' causes are forbidden).

- Atomic (fractionable).

- Reversible (no intrinsic temporal direction).

- Deterministic (given enough information about initial and boundary conditions, the future evolution of the system can be specified with any required precision).
- Universal (natural laws apply everywhere, at all times and scales).

Contemporary science shows that these claims are *all* false, in the literal sense that they work for some special kind of systems only (technically, they are not generic) (Depew & Weber, 1995; Adam & Groves, 2007; Ulanowicz, 2009; Louie & Poli, 2011). The framework that in many scientific quarters is presently under development includes open, non-fractionable, irreversible, non-deterministic and context-dependent systems.

While the traditional, reductionist strategy has proved enormously successful and cannot be simply abandoned, the problems that prove refractory to a reductionist treatment are growing, and this calls for the availability of complementary non-reductionist strategies. Reductionist methods work well when a system can be decomposed (fragmented) without losing information. On the other hand, for many systems, any fragmentation causes a loss of information (Poli, 2011b). The most promising alternative strategy is to substitute analysis via decomposition (the reductionist mantra) with analysis via natural levels (aka the theory of levels of reality), introduce indecomposable wholes and substitute Humean causation with powers and propensities. Note that, since indecomposable wholes are not (entirely) understandable from their parts, manipulation of parts may entail unexpected consequences (Popper, 1990; Rosen, 1985; Bhaskar,1988; Poli, 2010a,b, 2011a, 2012a,b; Louie & Poli, 2011).

Anticipatory Governance

Anticipatory governance is a system of prescriptions explicitly addressing the interplay among complexity, acceleration, and policy. Specifically, anticipatory governance provides a way to use foresight, networks, feedback and hierarchical links for the purpose of reducing risk and increasing the capacity to respond to events at their initial stages of development (Fuerth, 2011 2009; Quay, 2010; Louie & Poli, 2011). In order to deal with acceleration, organizations must acquire a greater sensitivity to weak signals and learn to respond them with increased flexibility and speed.

Anticipatory governance, as here understood, is an application of the theory of anticipatory systems to policy (Rosen, 1985; Poli, 2010a,c; Louie & Poli, 2011). Anticipatory methods include both traditional forecasting techniques (such as time series analyses, statistical elaborations and simulations) and more innovative foresight methods (including but not limited to Delphi, scenarios, and environmental scanning).

To start grasping some of the many subtleties of anticipatory systems, consider first an ordinary (i.e., nonanticipatory) dynamical system S. S may be an individual organism, an ecosystem, a social or economic system. Then, associate to S a second system, called a model M of S. The main condition that M has to respect is that the dynamic evolution of M runs faster than the dynamic evolution of S. In this way, M is able to predict the behavior of S. By looking at M, S obtains information about a later state of itself. Now, suppose that M and S can interact with each other, i.e., that M may affect S and S may affect M. The direction from S to M can be seen as an updating or an improving of M. This direction is rather straightforward and I shall omit its analysis. On the other hand, the opposite direction from the model M to the system S is much more intriguing. In order for M to affect S, M must be equipped with a set of effectors E, which allow M to operate on S (or on the environmental inputs to S) in such a way as to change the dynamics of S. If we consider S, M and E as parts of one single system, the latter will logically be an anticipatory system in which modeled future behaviors determine present states of the system.

A simple question will aid understanding of the connections among M, E and S: How can the information available in M be used to modify the properties of S through E? Consider partitioning the state space of S (and hence of M) into positive and negative states (from the point of view of S). As long as the dynamics of M remain in a positive region, no action is taken by M through the effectors E. When the dynamics of M move into a negative region (implying that the dynamics of S will later move into the corresponding negative region) the effectors are activated to keep the dynamics of S out of the undesirable region (for more details see Poli, 2010).

Formally speaking, S + M + E form a cycle connecting functions and their values (and vice versa). These cycles are called 'hierarchical' because 'function' and 'value' pertain to two different formal levels. Among the many properties of a natural system N that contains hierarchical cycles are the following (cfr. Louie 2009 for details):

1. N does not have a largest model (The largest model, if it exists, is the greatest element in the lattice of models, which implies that every model is its submodel).

2. Not every property of N is fractionable (A property of a natural system is fractionable if the natural system can be separated into two parts modeled by disjoint direct summands, such that the property is manifest in one of these parts).

3. There exist models of N that are not simulable (A model is simulable if every process is definable by an algorithm).

4. N is an impredicative system (Indeed, the containment of a hierarchical cycle may be used as a definition of impredicativity).

E:CO Vol. 14 No. 4 2012 pp. 124-138

Hierarchical cycle is used in the definitions of two important classes of systems: 'complex systems' and 'systems that are closed to efficient causation'.

Definition 1. *A natural system is complex if and only if it has a model that contains a hierarchical cycle (This definition of complexity is pretty different from mainstream acceptations of complexity; for preliminary discussions, see Rosen, 1985, 2000).*

Definition 2. *A natural system is closed to efficient causation if its every efficient cause is entailed within the system.*

Louie (2009) proves that the following two properties of a natural system are equivalent: (a) its every efficient cause is entailed within the system and (b) it has a model that has all its processes contained in hierarchical cycles. Stated otherwise, in a closed-to-efficient-cause system, all processes are involved in hierarchical cycles. Thus, the class of systems that are closed to efficient causation forms a proper subset of the class of complex systems (which are required to have only some processes involved in hierarchical cycles). Because of this containment, a closed-to-efficient-cause system may be considered a 'higher-order complex system'.

Instead of the verbose 'closed-to-efficient-cause system' or 'systems that are closed to efficient causation', Louie and Poli (2011) introduced the term 'clef system' (for closed to efficient causation) with the following definition.

Definition 3. *A natural system is clef if and only if it has a model that has all its processes contained in hierarchical cycles.*

There are different families of these clef systems that are of 'higher order' than complex systems. Three of them have been exemplified in Louie and Poli (2011), namely living, psychological, and social systems.

The difference between living, psychological, and social systems can be made more explicit by resorting to the theory of natural levels (or 'levels of reality'). The main distinction here is between 'material', 'psychological', and 'social' levels, the ontological categories that are embedded within them and the relations of dependence and independence among the different levels of reality (for details, see Poli, 2001, 2006a,b, 2007).

Two Types Of Strong Signals

The following two baby-exemplifications should help readers to distinguish between strong and weak signals. Firstly, consider one of those curves that we all studied when we were high-school students: the parabola. Parabolas have the surprising property of being entirely embedded within each of their points. From any point of a parabola one can reconstruct the entire figure simply by calculating the point's first and second derivatives. This situation is a good exemplification of a case of perfect local/global solidarity, the vision underlying Laplace's absolute determinism. If you know the position of an element and the forces acting on that element (recall that from a physical point of view, first and second derivative respectively mean velocity and acceleration), the future and past (i.e., the entire story) of that particle will be in plain sight with no surprise, no creativity, no spontaneity, no uncertainty, no difference between past and future. Everything is determined forever. One of the most intriguing aspects of absolute determinism is that the perfect local/global solidarity which it embeds may require only tiny amounts of information (in the given example, the first two derivatives). While most of classical science was grounded on a version of absolute determinism, contemporary science has shifted towards non-linear and chaotic systems, i.e., systems that breach local/global solidarity and which are such that even unlimited amounts of information may not allow forecasts to be made about the system's evolution.

Secondly, consider the climate and the models that scientists have developed to understand its change. Most models run until 2100 (and even beyond). The main reason for working with such an extended temporal window is that asymptotic foresights are usually far simpler than short-to-medium term ones. For instance, long-term foresights cancel out most of the 'noise', such as outliers and those occasional events that may be relevant in the short period but prove irrelevant in the long run. Climate models show that global phenomena take precedence over local ones. Therefore, climate models show a different way to breach the perfect local/global solidarity exhibited by analytic curves such as parabolas.

The assumption of a perfect local/global solidarity makes things simple: in most cases, small amounts of information are sufficient to obtain the results required.

Breaching local/global solidarity makes matters far less straightforward. Not only does more information become necessary, but even unlimited amounts of information may prove insufficient for any specific forecast. Furthermore, local/global solidarity may break down in two different ways: either the global takes precedence over the local (climate models) or the local takes precedence over the global (which eventually arises by 'gluing' together many local 'areas'). In the former case the local is not analytically calculable from the global; in the latter case the global is not analytically calculable from the local. Either way there are surprises, and synthetic models should flank analytic ones.

Weak Signals

Trends begin, grow by following more or less complex patterns, and eventually decline. Even if we dismiss the perfect-solidarity thesis, it is reasonable to assume that trends present some kind of pattern. The capacity to discern a wide variety of patterns is precisely what makes an expert, expert. The very beginning of a trend, however, is a critical point. Critical points can occasionally be seen from *within* trends (e.g., flection points, such as when a positive trend becomes negative or vice versa), but there is apparently no way to detect a singularity without a supporting trend. Weak signals (or seeds of change or early warning signals) are such a troublesome issue precisely because they are not part of an already established trend. They may eventually become the beginning of a trend. But when they occur, they are 'stateless' so to speak; the trend of which they will eventually become the beginning still does not exist.

The following three observations, by delving into progressively deeper waters, may aid understanding of weak signals.

Context

While the trend that eventually arises from a collection of weak signals still does not exist, weak signals themselves are elements of a wider context, a dynamic panorama with its habits, tensions, achieved results and unresolved issues.

Spontaneity

Even more relevantly, social contexts are based on different levels of depth characterized by a variety of principles. Two of the foremost principles are provided by the oppositions of 'visible vs. invisible' and 'constrained vs. spontaneous'. The underlying intuition is that phenomena at higher levels are more visible and more constrained than those at lower levels. On descending the hierarchy of levels, the corresponding social phenomena become more flexible and spontaneous, and it becomes more difficult for the social scientist to detect their presence and find their rules. Furthermore, the hierarchy of levels also appears to follow the principle according to which higher levels are always *partial* realizations of lower levels. In other words, lower levels are systematically richer, more nuanced, than higher levels. On the other hand, higher levels are concretizations, realizations of lower levels; they transform the latter into more 'visible' and 'objective' realities, and in doing so they somehow lose some of the richness and flexibility of the lower levels. The rock-bottom level is represented by spontaneous behavior and includes sparkling, innovative, creative behaviors (Gurvitch, 1964), some of which may be 'captured' by higher-order levels (values, symbols, attitudes, roles, procedural behaviors, etc.). In this way they both lose part of their flexibility and acquire more explicit degrees of visibility and objectivity.

Some comments are in order. (1) Levels of depth are not connected to levels of relevance or value. From the point of view of relevance of value, all the levels are on a par: no level is more relevant, important or valuable than any other. (2) Furthermore, levels of depth have different degrees of strength in different societies. The variable strength of levels of depth in different societies can be exploited as a tool with which to analyze the internal organization of entire societies. (3) The bottom line of the series interfaces with individual and collective minds. Here, Luhmann and Gurvitch come close to each other. Luhmann (1985) claims that the 'psychological system' should be seen as constituting the environment of the 'social system', and, at the same time, the 'social system' as constituting the environment of the 'psychological system'. Gurvitch (1964) says, on the other hand, that the psychological and the social are reciprocally immanent, apparently suggesting the same idea. (4) Levels of depth are mutually interconnected, they interpenetrate one another. Properly speaking, levels of depth are not separable in the sense in which things can be separated. They do not exist in a 'pure' or 'isolated' form. They exist only 'together', within the whole that is the overall social phenomenon. Levels can better be seen as 'foci' that may be analytically separated without being existentially separable. However, the co-penetration of levels of depth does not imply coherence or convergence. Levels create tension, they contend against one another in multifarious forms, and 'unresolved residues' may always result from their conflicts (Poli, 2010, 2011, 2012).

Causality

While levels of depth naturally provide a vertical framework of analysis, a temporal or dynamic framework is required as well. The main issue is causality as the connecting tie between the subsequent stages of real processes in which each stage produces the subsequent one, in a given temporal order. The idea here is that causality is *productive*, it generates the new subsequent stage. Production in this sense is not severed from creativity, which implies that the new stage is not included, embedded, or already contained in the preceding stage. For more details on the idea of causality broadly summarized here see the pertinent sections of Hartmann (1938, 1940, 1950). Before concluding, at least one aspect of Hartmann's theory of causality deserves to be explicitly underlined. By including creativity within causality, the general framework underlying determinism is demolished to its foundations. It is also worth noting that Hartmann's idea that causality is structurally creative comes close to Popper's theory of propensities (Popper, 1990). While the latter is definitely more well-known, Hartmann's theory both antedates Popper's by a few decades and is more explicitly worked out (for an introduction to Hartmann's ontology, see Poli, 2012a,b).

Apart from the first observation on context, which is pretty obvious, the other two observations, on levels of depth and causality, offer only a roughly sketched outline of the issues involved. Many more details are needed for these to become fully acceptable proposals. However, if it turns out that this preliminary sketch is even partially correct, weak signals will prove to be much more important than is usually believed, not least because they concern some of the deepest-lying intricacies of reality.

Entries For An Anticipatory Governance Dictionary

In what follows I shall add to the analysis of weak signals a series of aspects or components that may eventually merge into a coherent framework, if a theory of anticipatory governance is ever developed. I shall present them in the form of a network of mutually interlinked concepts—a kind of dynamic dictionary—including (1) Anticipatory governance and anticipatory systems;

(2) Regulation; (3) Resilience; and (4) Foresight maturity model. Let us consider them in turn.

Anticipatory Governance And Anticipatory Systems

While one of the main characteristics of human beings is their ability to foresee both their own development and that of the environment in—and on—which they operate (cf. the popular wisdom that distinguishes between "taking the long view" and "not looking beyond one's own nose"), institutional and organizational anticipations are characterized by a specific type of complexity. More often than not, the working style and other systemic features of institutions and organizations make them incapable of developing suitable anticipations, even if the individuals who populate them have good individual anticipatory capacities. After Rosen 1985 and 2000, the theory of anticipation has been widely discussed in the past few years. (For a summary see Poli, 2010a, for a comprehensive bibliography see Nadin, 2010.)

Nevertheless, the complexity of social and institutional anticipations—and in particular the issue of anticipatory governance—remains an underexplored topic (see, however, Fuerth, 2009; Quay, 2010).

The theory of anticipatory systems adds hierarchical cycles (aka impredicative cycles or self-referential cycles) to the theory of systems. The presence of hierarchical cycles dramatically constrains the modeling of the relevant system. To mention only one result, a system containing a hierarchical cycle must have a non-simulable model, which implies that no simulable description of that system will ever be complete. This result does not imply that there can be no model of hierarchical cycles at all. There are plenty of useful algorithmic models, but with the caveat that these will be, by definition, incomplete. They may nevertheless be fruitful endeavors. One learns a tremendous amount even from partial descriptions (Louie & Poli, 2011).

Regulation

Regulation primarily focuses on specific sectors or activities, and its major goal is to prevent accidents or disasters (the precautionary approach). The objective is proactive: reducing harm before risks emerge rather than making systems resilient to respond to them afterwards (for extensive examples of the weakness of the precautionary approach in the health sector, see Wismar *et al.*, 2007). Unfortunately, the process of developing and implementing regulations is often clumsy, fragmented, overly complicated and narrowly focused. As a consequence, regulation has trouble keeping up with the accelerated pace of global change (World Economic Forum, 2012: 22).

Therefore, a different approach to regulation is needed. The primarily *backward-looking* attitude that underlies the precautionary approach should be replaced with a different, *forward-looking* one, such as anticipatory governance. According to this approach, decision-makers accept that past experience alone is inadequate for generating a robust policy and that tools and methods for 'visualizing' the future are needed. Furthermore, it is acknowledged that safeguards cannot be established once and for all; they have to be continuously re-defined and negotiated. This implies that the systems to be safeguarded and the safeguards themselves evolve together through adaptive learning. The following are some of the main aspects of anticipatory governance applied to regulation:

- Evolutionary perspective—safeguards are unlikely to be right the first time they are implemented.

- Iterative and incremental adaptive learning process between regulators and experts at the frontiers of knowledge.

- Search for and systematic adoption of best practices (see below).

- Avoiding situations in which one error could be catastrophic.

- Real-time monitoring.

- Collaboration and the sharing of knowledge—authorities that define safeguards must also tell people they have done so.

Resilience And The Sense And Sensibilities Of Coherence (SSOC) Index

Will the processes currently in use lead society towards a more desirable future or will they create new threats and more instability? According to the seventh edition of the report on *Global Risks* (World Economic Forum 2012), the linkages among a variety of global risks—including natural, demographic, fiscal and societal ones—call attention to the emergence of a new class of *critically fragile countries*—or regions, for that matter. The report shifts its focus from understanding *individual* types of risk to understanding their *systemic* connections. Failures in managing aging populations, youth unemployment, growing inequalities and fiscal imbalances may increase the dangers of social unrest and instability in the years to come.

Following Bennett 1996, anticipation is seen as the basis of human adaptation. The question then arises as to which features or properties make communities more likely systematically to develop anticipatory understanding and behavior. A key application of anticipatory studies is the analysis of current dynamics and trends in society in order to facilitate the taking of decisions oriented to the long term and make the social system more resilient to undesirable change. (On the various levels of resilience, see Martin-Breen & Anderies, 2011.)

Most discussions on resilience fail to draw an explicit link between resilience and anticipation. Almedom (2007, 2009) is a rare exception. She has proposed the following definition: *Resilience is a multidimensional construct defined as the capacity of individuals, families, communities, and institutions to anticipate, withstand and/or judiciously engage with catastrophic events and/or experiences; ac-*

tively making meaning out of adversity, with the goal of maintaining 'normal' function without fundamental loss of identity. This definition highlights that people (at both individual and collective levels) and their formal and informal institutions are crucial for the sustainability of coupled social-ecological systems.

The Sense and Sensibilities of Coherence scale (SSOC) measures the capacity of people and communities to manage stressful situations and remain healthy (Almedom *et al.*, 2007; Tiberi *et al.*, 2012). The SSOC takes into account multiple variables that constitute the socio-cultural, historical, and geo-political context in order to gauge human resilience at the societal level. The SSOC will document adaptive learning that facilitates the successful management of the critical transition towards transformation that reflects adaptive governance.

Foresight Maturity Model

In order for the anticipatory exercise to help improve individual and collective choices, the future of the individual, community, business, organization or institution that decides must become a continued, common and shared subject of discussion and exchange of ideas. If the future of the deciding unit itself is to become a force that contributes to the steering of communities, businesses, organizations or institutions, then it must be discussed explicitly.

After more than sixty years' experience, the critical factors that normally determine the success or failure of institutional projects of social forecasting are now well known (Calof & Smith, 2010). The key factor is whether or not present mental and social models of the community are prepared for future challenges. In order to change social models within a community, one needs vision, strategy, and systematic monitoring of changes as they happen.

Vision involves reference values: why should we wish to change the currently active patterns? A few answers, in order of 'transparency', are the following: so as to increase the community's social capital, to improve its sustainability, to develop a community with a higher level of resilience. Strategy is about how to proceed, and monitoring will keep track of the evolution of the project and verify whether the (intermediate and final) objectives have been achieved. It will be necessary to adopt a number of indicators of the cultural and behavioral changes and the level of resilience of the system—and possibly develop new ones. Another matter of importance is whether the results are robust: will the new models remain stable over time?

The interactive process of collective intelligence and the mobilization of the players can create, in itself, consensus on and commitment to the action that will be undertaken, especially when identifying common stakes and developing a shared vision. Additionally, foresight should lead to a strategic phase that will give precise answers to the long-term challenges identified and to an action program potentially capable of attaining that vision. To be credible, the program will have to: 1) be as explicit as possible; 2) address the key constraints identified; 3) identify the actors and the means to carry out the actions; and 4) pay especial attention to budgetary choices. The action program will also have to include guidance for the implementation and follow-up, along with the process's evaluation and exercise's products.

The means of assessing an organization's capability has been developed as the Foresight Maturity Model (FMM) (Grim, 2009). FMM is part of a family of rubrics used to assess institutional capabilities in various fields. It was first implemented in software engineering so that the US Department of Defense could assess the capabilities of contractors to deliver correct software on time and within budget. The main task of FMM is to assess how well organizations are practicing foresight by relying on examples of 'best practices' in their field. While the concept of 'best practice' is often elusive, FMM provides the analytical tools for strengthening foresight. (On the problems surrounding 'best practice', see Coote *et al.*, 2004; Auspos & Kubish, 2004; Foot *et al.*, 2011.) The two main features of FMM are (1) devising the main steps of a foresight exercise, as developed by Hines and Bishop (2006) and (2) identifying the series of basic maturity levels. FMM organizes a foresight exercise into six different components, and five different maturity levels are distinguished for each of them (Table 1).

Disciplines of a Foresight Exercise	Basic Maturity Levels for Each Discipline
Leadership (capacity to translate foresight into action on an ongoing basis)	**Ad Hoc** (being only marginally aware of processes; most work is done without plans or expertise)
Framing (capacity to identify and solve the right problems)	**Aware** (being aware that there are best practices in the field; learning from external input and past experiences)
Scanning (capacity to understand what's going on in the immediate environment and in the world at large)	**Capable** (having a consistent approach for a practice, used across the organization, which delivers an acceptable level of performance and return on investment)
Forecasting (capacity to consider a range of future possibilities)	**Mature** (investing additional resources to develop expertise and advanced processes for a practice)
Visioning (capacity to decide what the organization wants in the future)	**World-class** (being a leader in its field, often creating and disseminating new methods)
Planning (capacity to develop plans, skills, and processes that support the organization's vision	

Table 1 *The main components of the Foresight Maturity Model*

E:CO Vol. 14 No. 4 2012 pp. 124-138

Conclusion

Providing that the above-sketched proposal is at least partially correct, the following are some of the conclusions that may be inferred from it:

- If we embrace complexity, the difference between natural and social systems acquires a different flavor. In short, while physical systems are primarily reactive, living and social ones are primarily anticipatory. Different forms of anticipation may be detected within non-physical systems: while the anticipatory capacities of a cell lie below the threshold of awareness, complex organisms or collectives exhibit cognitive (i.e., explicit) anticipatory capacities.

- "Propensities" (to use Popper's term) tend to be low in the physical realm and high in the social one. With reference to propensities as weak signals, there are almost nil weak signals in the physical realm, and boundless weak signals in the psychological and social realms.

- It is advisable to avoid the trap of the "either-or" assumption. Questions of the type "are weak signals independent or culturally dependent?" are bound to be wrong from the outset because both physically independent weak signals and psychologically or socially dependent ones can be detected. Furthermore, at least some weak signals may arise from the interplay between the physical and the cultural (a stone is a stone, but when I move it somewhere else, it is no longer an *exclusively* physical something).

- While different people and cultures may be more or less able to detect weak signals, institutions appear to be structurally blind to most weak signals. If this is the case, there is a difference between the capacity to detect weak signals exhibited by persons and informal groups and those exhibited by organizations and institutions.

Therefore, how can one make an organization or institution systematically open to (at least some kinds of) weak signals? Here is my proposal:

- Ground your framework in the theory of anticipatory systems, which is presently the most general framework for understanding self-referential or impredicative phenomena. Anticipatory systems are characterized by the presence of hierarchical cycles and have properties remarkably different from those of systems without hierarchical cycles. In short: they do not have maximal models, which implies that they contain at least one non-simulable model.

- Understand that past experience alone is not enough to generate a robust policy and that tools and methods for 'visualizing' the future are needed. Furthermore, safeguards cannot be established once and for all; they have to be continuously re-defined and negotiated. This implies that the systems to be safeguarded and the safeguards themselves evolve together through adaptive learning.

- Combine resilience and anticipation. While much work is still needed to acquire a robust understanding of the proper way to join resilience and anticipation, the Sense and Sensibilities of Coherence scale (SSOC) is a preliminary effort to measure the capacity of people and communities to manage stressful situations and remain healthy.

- Assess how well organizations are practicing foresight. The Foresight Maturity Model is presently the most advanced framework for identifying the series of basic foresight maturity levels.

References

Adam, B. and Groves, C. (2007). *Future Matters*, ISBN 9789004161771.

Almedom, A.M. (2009). "A call for a resilience index for health and social systems in Africa," *Issues in Brief*, 10, Boston University, Boston, Frederick S. Pardee Center for the Study of the Longer-range Future.

Almedom, A.M., Tesfamichael, B., Mohammed, Z.S., Mascie-Taylor, C.G.N., and Alemu, Z. (2007). "Use of 'sense of coherence (SOC)' scale to measure resilience in Eritrea: Interrogating both the data and the scale," *Journal of Biosocial Science*, ISSN 1469-7599, 39: 91-107.

Auspos, P. and Kubisch, A.C. (2004). *Building Knowledge about Community Change*, The Aspen Institute.

Bennett, J.W. (1996). *Human Ecology as Human Behavior: Essays in Environmental and Development Anthropology*, ISBN 9781560000686.

Bhaskar, R. (1988). *The Possibility of Naturalism*, ISBN 9780415198738.

Calof, J. and Smith, J.E. (2010). "Critical success factors for government-led foresight," *Science and Public Policy*, ISSN 0302-3427, 37: 31-40.

Coote, A., Allen, J and Woodhead, D. (2004). *Finding Out What Works: Building Knowledge about Complex, Community-Based Initiatives*, ISBN 9781857174861.

Depew, D.J. and Weber, B.H. (1995). *Darwinism Evolving: System Dynamics and the Genealogy of Natural Selection*, ISBN 9780262041454.

Foot, C., Raleigh, V., Ross, S. and Lyscom, T. (2011). How do Quality Accounts Measure Up? Findings from the First Year, ISBN 9781857176087.

Fuerth, L. (2011). "Operationalizing anticipatory governance," Prism, ISSN 2157-0663, 2: 31-46, http://www.ndu.edu/press/anticipatory-governance.html.

Fuerth, L.S. (2009). "Foresight and anticipatory governance," *Foresight*, ISSN 1463-6689, 11: 14-32.

Grim, T. (2009). "Foresight Maturity Model (FMM): Achieving best practices in the foresight field," *Journal of Futures Studies*, 13(4): 69-80, http://www.jfs.tku.edu.tw/13-4/AE05.pdf.

Gurvitch, G. (1964). *The Spectrum of Social Time*, ISBN 9789027700063.

Hartmann, N. (1938). *Möglichkeit und Wircklichkeit*, ISBN 9783110001518.

Hartmann, N. (1940). *Der Aufbau der realen Welt: Grundriss der allgemeinen Kategorienlehre*, ISBN 9783110001471.

Hartmann, N. (1950). *Philosophie der Natur. Abriss der speziellen Kategorienlehre*, ISBN 9783110047493.

Hines, A. and Bishop, P. (2006). *Thinking about the Future: Guidelines for Strategic Foresight*, ISBN 9780978931704.

Louie, A.H. and Poli, R. (2011). "The spread of hierarchical cycles," *International Journal of General Systems*, ISSN 0308-1079, 40: 237-261.

Luhmann, N. (1985). *Social Systems*, ISBN 9780804726252.

Martin-Breen, P. and Anderies, J.M. (2011). "Resilience: A literature review," The Rockfeller Foundation, http://www.rockefellerfoundation.org/news/publications/resilience-literature-review.

Nadin, M. (2010). "Annotated bibliography: Anticipation," *International Journal of General Systems*, ISSN 0308-1079, 39(1): 5-133.

Poli, R. (2010a). "The many aspects of anticipation," *Foresight*, ISSN 1463-6689, 12: 7-17.

Poli, R. (2010b). "The Complexity of self-reference: A critical evaluation of Luhmann's theory of social systems," *Journal of Sociocybernetics*, ISSN 1607-8667, 8(1-2): 1-23.

Poli, R. (2010c). "An introduction to the ontology of anticipation," *Futures*, ISSN 0016-3287, 42(7): 769-776.

Poli, R. (2011a). "Steps toward an explicit ontology of the future," *Journal of Futures Studies*, 16(1): 67-77, http://www.jfs.tku.edu.tw/16-1/A04.pdf.

Poli, R. (2011b). "Analysis, synthesis," in V. Petrov (ed.), Ontological Landscapes: Recent Thought on Conceptual Interfaces Between Science and Philosophy, ISBN 9783868381078, p. 19-42.

Poli, R. (2012a). "Nicolai Hartmann", in E. Zalta (ed.), *The Stanford Encyclopedia of Philosophy*, http://plato.stanford.edu/entries/nicolai-hartmann/.

Poli, R. (2012b). "The difference between dynamical systems and process theories," submitted.

Popper, K.R. (1990). *A World of Propensities*, ISBN 9781855060005.

Quay, R. (2010). "Anticipatory governance," *Journal of the American Planning Association*, ISSN 0194-4363, 76: 496-511.

Rosen R. (2000). *Essays on Life Itself*, ISBN 9780231105118.

Rosen, R. (2012). *Anticipatory Systems: Philosophical, Mathematical and Methodological Foundations*, ISBN 9781461412687.

Tiberi, L.J., Parker, J., Akhilgova, J., Toirov, F., and Almedom, A. (2012). "'Hope is the engine of life'; 'Hope dies with the person': Analysis of meaning making in FAO-supported North Caucasus Communities using the 'Sense and Sensibilities of Coherence' (SSOC) Methodology," *Journal of Loss & Trauma*, ISSN 1532-5024, in press.

Ulanowicz, R.E. (2009). *A Third Window: Natural Life beyond Newton and Darwin*, ISBN 9781599471549.

Wismar, M., Blau, J., Ernst, K., and Figueras, J. (2007). *The Effectiveness of Health Impact Assessment: Scope and Limitations of Supporting Decision-Making in Europe*, ISBN 9789289072953.

World Economic Forum (2012). *Global risks 2012: Insight Report. An Initiative of the Risk Response Network*, ISBN 9789295044357.

The research interests of **Roberto Poli** (Ph.D. Utrecht) include (1) anticipatory systems and futures studies, (2) ontology, and (3) the theory of values and the category of person. Poli is editor-in-chief of Axiomathes (Springer). Poli has published six books, edited or coedited more than 20 books or journal's special issues and published more than 150 scientific papers. Poli teaches Social Foresight, Applied Ethics, and Philosophy at the Faculty of Sociology of the University of Trento (Italy).

Classical

Classic Paper Section

Life's Irreducible Structure

Michael Polanyi (with an introduction by Jeffrey Goldstein)

Originally published as Polanyi, M. (1968). "Life's irreducible structure," *Science*, 160(3834): 1308-1312. Reprinted with kind permission.

By shifting our attention, we may sometimes change a boundary from one type to another.

Polanyi (p. 1308)

Polanyi's Anti-reductionist Philosophy of Biology
Introduction

During the development of the sciences of complex systems certain thinkers stand out as much for the multi-thronged influences, direct and indirect, initiated by their work as for the content of their thought per se. Such was the case of the physician, chemist, biologist, social thinker, and philosopher Michael Polanyi (1891-1976) who, though born in Hungary, did most of his research and thinking first in Germany and then England (the biographical information found here is culled from: Nye, 2002; Polanyi A, B and C).

Polanyi's career followed a trajectory like that of other Hungarian polymaths, a prototype of intellectual achievement which John von Neumann attributed to the complex intellectual environment of Hungary as one of the two centers of the Austro-Hungarian Empire from the nineteenth century up to World War I. Most pertinent in this regard was the learning of several quite disparate languages including Hungarian with its unique and idiosyncratic Uralic-Altaic origins, German and that other Germanic language English, French and related Romance languages, and Slavic languages such as Czech, Polish, and Russian. For a child living in such a stimulating milieu and mastering such diverse tongues, it stands to reason they would develop a kind of cognitive complexity not come by having to lean just one language during their formative years. Like von Neumann himself, Polanyi was in grand company with other Hungarian intellectual giants like the physicists and mathematicians Wigner, Szilard, Teller, Erdos, Halmos, Bolyai, Bott, Erdelyi, Grossman (Einstein's friend and collaborator), Oskar

Klein, Lakatos, Rado, Polya, Renyi, Gabor, Barabasi, Lax, and many more—and all hailing from a small country whose population in 1949 was only 9 million (of course much diminished due to the ravages of WWII).

Polanyi's intellectual achievements were many and far-ranging. For instance, in 1930, in collaboration with the great physical chemist Fritz London, Polanyi offered a new and generalized explanation for adsorption forces (as dispersion forces) using the newly formulated quantum mechanics. Another focus of his work was x-ray diffraction studies of natural fibers and metals, developing the experimental method of chemiluminescence and 'highly dilute flames' jointly with Eugene Wigner and H. Pelzer. What a family life it must have been with Polanyi's eminent economist brother Karl and Polanyi's son John going on to win the Nobel Prize in Chemistry.

Polanyi traveled in erudite and influential circles indeed. He struck an intellectual friendship with Bukharin around the issue of the infamous Lysenko scandal involving Lamarkianism in Russia. From these and other experiences, Polanyi became increasingly interested in social dynamics. He was one of the first propounders of the idea of a "spontaneous order" arising in social systems, operating, for example, in free market dynamics, an early form of social "self-organization" that was taken-up in the idea of catallaxy, which Hayek made into a cornerstone of his vastly influential market-driven economic theory so important for instance in later work of Milton Freidman.

Polanyi came up with a very unique understanding of how science functioned, including how new discoveries occurred within scientific communities of practice. According to Nye (2002), Polanyi's singular idea of science as a cultural/social activity consisting of commitment to established dogmas was broached at a conference at Oxford in 1961 where Thomas Kuhn had coincidently summarized his thesis on paradigms, normal science, and scientific revolutions which would later be published as *The Structure of Scientific Revolutions*. For Polanyi, scientific knowledge is forthcoming within a social system of authority and apprenticeship in which scientific discovery has as much to do with the community of practice's current interests and presuppositions as it does with experimental or logical plausibilities. Polanyi's thoughts about the social milieu of science found an eager audience with no less a intellectual luminary than Alan Turing, a friendship about which more will be said later.

At the same time, Polanyi was foraging in other sciences beyond his home-ground of chemistry, particularly in theoretical biology. Here, Polanyi was elaborating on a sort of neo-teleological position which had been much effected by his studies of the work of the modern vitalist Hans Driesch. Polanyi was aware that he was walking on a kind of intellectual tight rope by taking vitalism's teleological principles seriously and accordingly he strived to persistently update these speculations in the light of advances in biological findings, particularly the discovery of DNA.

Polanyi's specific insights into biology and human life as complex systems became noteworthy enough in the early nineteen fifties that he was invited to deliver the celebrated Gifford Lectures in Scotland form 1951-52. By taking-up that honor, Polanyi stood in the same venerable precursor and early emergentist tradition of Henri Bergson, Wolfgang Kohler, John Dewey, William James, Conwy Lloyd Morgan, Samuel Alexander, Alfred North Whitehead, Conrad Hal Waddington, among others.

In his philosophy of biology Polanyi wanted to carve out an inviolable place for the study of life by demonstrating why the fundamental principles by which living organisms operate would in the last analysis always outstrip encroachments by reductive explanatory strategies. In other words, he hoped to show that no matter how well such disciplines as physics and chemistry might explain "lower" level life processes, these approaches could not ultimately explain what Polanyi held as the non-formalizable, irreducible overall principle of ordering which guided living organisms. Polanyi's classic paper, "Life's Irreducible Structure" employed his considerable creative ability to shift among various fields applying insights concepts and metaphors from one to another, political economy to philosophy, to theories of the mind, to molecular biology, and so forth.

One more intellectual accomplishment should be mentioned here for the bearing it has on the issue of Polanyi's anti-reductionist stance by way of his claims of non-formalizability. This was his aforementioned with the inimitable Alan Turing. As we'll say more about below, Turing was influenced by Polanyi's dedication to a non-formalizable understanding of morphogenesis. Albeit Turing was inspired by these conversations in order to counter Polanyi's arguments by seeking to devise a formalizable account, in terms of mathematics and chemistry, of precisely how morphogenesis could proceed entirely naturalistically and without the need to resort to any kind of "immaterialist" "higher" ordering principle. Turing's (1952) work in this area was, surely one of his most brilliant and lasting contributions to the sciences of complex systems.

Boundary Conditions, "Ordering Principle", and Irreducibility

On page 160 of his classic article, Polanyi wrote: "In this light the organism is shown to be, like a machine, a system which works according to two different principles: its structure serves as a boundary condition harnessing the physical-chemical processes by which its organs perform their functions... Morphogenesis, the process by which the structure of living beings develops, can then be likened to the shaping of a machine which will act as a boundary for the laws of inanimate nature. For just as these laws serve the machine, so they serve also the developed organism." We can see here that Polanyi couched his idea of an irreducible structure of life by resorting to the metaphor of a machine, a somewhat ironic explanatory strategy given how anti-reductionist arguments usually proceeded by emphasizing some specifically *non*-mechanical foundation to complex systems. What Polanyi was trying to accomplish with his machine analogy, however, was to apply the idea of a machine's boundary condition (stemming from what we can call the "meta"-level of the machine's designers and thereby determining the activities of the machine's parts and how these parts interact with each other) to a cognate boundary condition role for the structure of a living organism. And just as a machine's meta-level boundary condition of design cannot be reduced to the level of the machine's parts so the meta-level *structure* of an organism must be recalcitrant to being reduced to the "lower" level processes harnessed by the structure.

By "structure" of the living organism Polanyi's was expressing his teleological point of view of life which he had first taken-up from his studies of the vitalist Hans Driesch's. With the discovery of DNA, Polanyi now had a concrete, naturalistically based support for his teleology, a blueprint putatively guiding morphogenetic development, all physiological parts and processes allied toward the telos of the organism by way of the information contained within DNA. What better instance of a boundary condition harnessing "lower" level processes could Polanyi have wished for? And, furthermore, DNA and its linear sequences of bases were thought to arise through a long evolutionary history of being subject to selective forces and, in that way at least, became a boundary condition not vulnerable to reduction by physics or chemistry.

A persistent temptation dogging advocates of emergence has consisted of supporting the idea of a purported irreducibility of "higher" level wholes through postulating some kind of factor, force, principle, structure, "form", field, law, or entity which one way or another, because of its transcendence of the merely natural, could insure against naturalistically-framed explanations aiming at reduction. The transcending power of such a force is one of the reasons emergence has been embraced by many within a theological or spiritual fold. Because it is conceived as something not thought to be built out of system components, this "higher" level force is believed to not be subject to an explanation entirely in terms of those components. Moreover, this "higher" level irreducibility offers the benefits of accounting for how "lower" level parts cohere or are integrated in emergent phenomena and thus how this integration could possess the property of downward causation which so many emergentists have claimed is central to their doctrine.

Yet, among the conceptual challenges incumbent upon positing such an irreducible "higher" level, one issue stands out as particularly crucial for any approach to emergence that aims at being scientifically viable, namely, how to envision it without invoking the concern that what is being appealed to cannot be approached scientifically. Polanyi in fact gave- in to just this temptation with his description of an "ordering principle" understood in a vitalist cast in order to answer his questions: "How can the emergent have arisen from particulars that cannot constitute it? Does some new creative agent enter the emergent system

E:CO Vol. 14 No. 4 2012 pp. 139-153

at every new stage" (Polanyi, 1958: 393). Polanyi credited the modern re-discovery of this "ontogenetic" principle of "morphogenetic regulation" to Driesch's work with the embryo of the sea urchin Clayton (2008). As Clayton emphasizes, Polanyi was adamant that the action of such a principle was not formalizable in terms of anatomical "machinery".

Indeed, Polanyi (1958: 423) was kind of hung-up on his notion of non-formalizability (another way for him to claim "not reducible to"):

> ...I believe that the unformalizable regulative functions, linked to the animal's mental processes, are the predominant, comprehensive agency of animal life....that evolution can give rise to ever new unformalizable operations only by acting itself, as an unformalizable principle—new machine-like operations can likewise emerge only in the same unformalizable manner.

Of course, it is ironic that Polanyi appealed to machines, the prototype of formalizability, as the prototype of the non-formalizable. The key for Polanyi in performing this ironic twist was to posit an ultimate irreducibility of the machine because its design comes from "outside" it, so to speak.

Polanyi did concede his "ordering principle" had not yet been established by science, but he also held that it would be as science advanced. It is interesting to note that it was in part a repudiation of the immaterialist connotations of Polanyi's "ordering principle" that Turing devised his brilliant mathematico-chemical conceptualization of the emergence of novel order seen in morphogenesis (Hodges, 2012). Turing (1952) showed just the opposite, namely, how to think in purely naturalistic means about phenomena that seemingly prompted a non-naturalistic explanation (Roth, 2011; also see Leiber, 2000).

Indeed, not only had Polanyi's time at Manchester overlapped that of Turing, but also his interests and proficiencies. An example was Polanyi's Gödelian argument to the effect that the mind had capacities beyond any merely mechanical system, a thesis that he used in his claim that the former could not be reduced to the latter. I find it quite interesting and telling about his own polymathic abilities that Polanyi had delved into Gödel's work in such a deep fashion way before it later moved to the forefront intellectually in the work of the English philosopher John Lucas and still later in the controversial speculations put forward by the renowned English mathematician and physicist Roger Penrose.

Turing was not comfortable with the particular thrust of Polanyi's Gödelian arguments about the mind, offering instead penetrating computationally-based insights into this issue as he pondered on the relation of mentation to artificial intelligence and mathematical logic (Hodges, 2012). Of course, the latter two conceptual arenas were show cases for Turing's inimitable genius demonstrated both in his extension of Gödel's Incompleteness Theorems in his work on Non-computable Numbers and his ground-breaking forays into computation when he intellectually spearheaded the effort to break the Enigma Code at Bletchley Park during WWII.

Polanyi (1958) discussed the functioning of his "ordering principle" in the language of a stable, self-sustaining, open system along the lines of a flame described by the great chemist Friedrich Wilhelm Ostwald (after all, Polanyi's main career was that of a chemist): if a flame is ignited accidentally by friction between combustible materials and if there are enough combustible materials present and if there is the condition of an "open" system meaning a continuous input of combustible gases and a capacity to transfer heat to an environment, the flame will become both stable and self-sustaining. A fundamental property of such stable open systems, according to Polanyi, is its stabilization of those fluctuations which first elicit them. But, in explaining how a flame could continue to exist, Polanyi realized it was not sufficient to appeal only to its initial, triggering ignition but also required the inclusion of an "ordering principle" operative in its self-sustaining nature. The trigger only served to "release" what is pre-given and it is this pre-given principle that prompted consecutive novelties which gradually achieve dominance over the course of evolution, "Novel forms of existence take control of the system by a process of *maturation*" (his emphasis).

Moreover, in couching his "ordering principle" in just these terms, we can appreciate Polanyi's concept as providing a way to incorporate continual epigenetic and not just predetermined evolutionary development. In an analogy offered by the contemporary theologian John Haught (1980) who, taking off on a suggestion about emergent levels offered by the eminent philosopher Marjorie Grene, an architect's designs do not interrupt the continuity of brick laying but rather impose a particular structure, from the design, on the layout of the bricks. Since this structure is one of the boundary conditions, the implication is that the construction of the building introduces a discontinuity among levels (in Polanyi's sense).

Conclusion: "Life Itself" Is Most Itself When it is Not Itself

The philosopher of biology Roger Faber (1986) has pointed to an arbitrariness in the two-fold distinction Polanyi assumed for boundary conditions. Polanyi conceived one kind along the lines of something like the shape of a pan impressing itself upon a cake being baked in it; the second was like the design that fixes the patterns of action found in a machine. Whereas the shape-bearing pan could be reduced to a deformation of the atoms and molecules of the pan's material, the second resists reduction since it is on a "higher", "meta-" level, than the machine components themselves. This second kind of boundary condition functions as a "higher" level structure harnessing the machine's "lower" level parts properties towards its own purposes. As we have seen, according to Polanyi, a living organism's structure was just this type of "meta-" level design which, although organizing the multitude of "lower" level processes into a viable integration, was not accessible from a "lower" level reductive understanding.

The philosopher Robert Causey (1969) claimed the arbitrariness on Polanyi's part implied that the second type of boundary condition could in fact be shifted into the first kind, thereby rendering it reducible. In other words, the ordering principle itself, operative in the boundary condition of design, could similarly be transmuted into the first kind and thus be subjected to reduction. Causey's argument appears to me to be quite similar in intention and outcome to the philosopher Jaegwon Kim's (1993) contention that emergence can be understood as a kind of supervenience and that as such all emergent phenomena are ultimately of the same ontological status as the components from which they are made, and thus capable of being reduced to just these components.

It seems to me, however, that both Causey and Kim have failed to recognize a crucial fact about complex systems, namely, that there are factors and forces at work which serve to so radically transform system components to such a degree that the principles of their functioning no longer can be explained sufficiently through a reduction to them as they were before this radical transformation (in this context, see Humphries, 2007, on processes of "fusion" during emergence which may change system parts so thoroughly they are no longer present as such for the "higher" level to be reduced to them) . What I am getting at here with the idea of a radical transformation are things like: the effects of contingencies and random events which form DNA's sequences out of a long chain of historical happenstance (with effects like the pan's shape in the metaphor used above); nonlinearities in the system which may contain "jumps" of system properties associated with the emergence of new attractors; and the "multifoldedness" (Ehresmann & Vanbremeersch, 2007) of "lower" level entities which allows them to be combined into radically complex new forms.

What Polanyi was specifically getting at with his idea of *structure* can be clarified by comparing it to the idea of *organization* which functions as the conceptual linchpin in Francisco Varela's notion of autopoeisis. Ostolaza and Bergareche (Helmreich, 2000), highlight the critical difference between *organization* (Polanyi's *structure*) and what Varela termed "*structure*" (not Polanyi's meaning of what he termed "structure"): "organization" referred to that whose objective is the preservation of itself, i.e., the self-referentially circumscribed nature of autopoeitic closure whereas "structure" was defined as the material substrate which is organized by the organization for the purpose of survival. Like Varela's *organization*, Polanyi's *structure* harnesses the "lower" level processes so that they conform to the putatively irreducible design and purpose of the organism. In the work of both Polanyi and Varela, structure or organization is conceived as independent from the causally understood processes at work on the "lower" level (see Rosen's "closed to efficient causality" idea below). And, it is this independence from the "lower" which renders both structure and organization irreducible.

A similar explanatory strategy within the philosophy of biology can be found in the work of Robert Rosen (see, e.g., *Life Itself*, 2007) with his concept of Metabolism/Repair (M/R) Systems, which Rosen argued were not reducible due in part by being "closed to efficient causation". All three philosophies of biology, that of Polanyi, Rosen, and Varela sought to create an almost sacred arena for studying life which would be inviolable to encroachment by the reductive explanations of physics and chemistry. Although these explanatory moves did serve to emphasize irreducibility in the complex system of living organisms, they also, in my opinion, sowed the seeds of an irreparable deficiency in their respective schemes which I will describe here.

The basic flaw stemmed from the diremption of their theories from the actual lived life of the living. As the systems thinker Rod Swenson (1992) remarked about the theory of autopoeisis, "[It] is miraculously decoupled from the physical world by its progenitors...[and thus] grounded on a solipsistic foundation that flies in the face of both common sense and scientific knowledge". Moreover, Swenson further indicated that one of the formulators of autopoeisis, Humberto Maturana, refused to sanction the use of the term "self-organization" as a synonym for autopoeisis since he held it implied that such a system was capable of change, whereas autpoeitic systems are fundamentally invariant in their organization. And Rosen (1970; p. 12) himself wrote,

> We have neglected developmental problems; and most particularly, we have neglected evolutionary problems, which are concerned with the way in which the class of organisms changes over long periods. Since a "structural" analysis pertains only to the class of biological systems at single instants of time (i.e. is a static description of the biological world, considered in evolutionary terms), there is an essential dynamical element missing from our discussion; in the terms used above, we have specified the instantaneous states of the biological world, but not the forces acting on them to produce changes of state, nor the equations of motion to which these forces give rise.

But it is not just change and development and evolution that these "higher" level philosophies of biology disregarded—it was also reproduction, sexual relations, sexual ecstasy, embeddedness in family, group, and/or social milieu, and a capacity for being effected by "lower" level fluctuations as seeds of evolutionary novelty. Indeed these three approaches amount to an egregious elision of variation, transformation, and transcendence altogether. What we are left with instead is an abstraction, self-enclosed, an essentialist rendering of life revealed in Rosen's telling phrase "life itself". But isn't life most itself when it is precisely not itself? When it is rather transcending itself in reproduction and family, in sexual ecstasy, in evolutionary transformation?

I have found it exceedingly strange, to say the least, that all three conceptual-izations of life—Polanyi's structure, Varela's autopoeisis, and Rosen's "closed to efficient causation" M/R systems—glaringly left out central factors defining life as life. It all seems like a sort of creepy neo-Victorianism. Perhaps related to this has been something else I've noticed among the extreme acolytes of these per-suasions, a sort of glassy-eyed defensiveness at what they take as slights aimed at their master's words. I don't presume to know what is psychologically at work here. But, I just don't see how this kind of Kool-Aid is of any real use in the study of complex systems,

References

Causey, R. (1969). "Polanyi on structure and reduction," *Synthese*, ISSN 0039-7857, 20(2):230-237.

Clayton, P. (2008). "Conceptual foundations of emergence theory," in P. Clayton and P. Davies (eds.), *The Re-Emergence of Emergence: The Emergentist Hypothesis from Sci-ence to Religion*, ISBN 9780199544318.

Ehresmann, A.C. and Vanbremeersch, J-P. (2007). *Memory Evolutive Systems: Hierarchy, Emergence, Cognition*, ISBN 9780444522443.

Faber, R. (1986). *Clockwork Garden: On the Mechanistic Reduction of Living Things*, ISBN 9780870235214.

Goldstein, J. (2003). "The construction of emergent order, or how to resist the tempta-tion of hylozoism," *Nonlinear Dynamics, Psychology, and Life Sciences*, ISSN 1090-0578, 7(4): 295-314.

Haught, J. (1980). *Nature and Purpose*, ISBN 9780819112583.

Helmreich, S. (2000). *Silicon Second Nature: Culturing Artificial Life in a Digital World*, ISBN 9780520208001.

Hodges, A. (2012). *Alan Turing: The Enigma*, ISBN 9780691155647.

Hubert, C. (nd). "Mechanism/Vitalism," http://www.christianhubert.com/writings/mech-anism_vitalism.html.

Humphreys, P. (2007). "How properties emerge," in M. Bedau and P. Humphreys (eds.), *Emergence: Contemporary Readings in Philosophy and Science*, ISBN 9780262524759.

Kim, J. (1993). *Supervenience and Mind: Selected Philosophical Essays*, ISBN 9780521439961.

Leiber, J. (2000). "The biological tradition of D'Arcy Thompson and Alan Turing," http://mailer.fsu.edu/~jleiber/D%27ArcyTuringChomsky.htm.

Nye, M. J. (2002). "Hyle Biographies: Michael Polanyi (1891-1976)," *HYLE: International Journal for Philosophy of Chemistry*, ISSN 1433-5158, 8(2): 123-127.

Polanyi A. "Michael Polanyi," Wikipedia, http://en.wikipedia.org/wiki/Michael_Polanyi.

Polanyi B. "Gifford Lectures Biography: Michael Polanyi," http://www.giffordlectures.org/Author.asp?AuthorID=139.

Polanyi C. "Michael Polanyi, FRS". R.T. Allen, The Society for Post-Critical and Personalist Studies (UK), http://www.spcps.org.uk/polanyi.pdf.

Polanyi, M. (1946). Science, Faith and Society, ISBN 9780226672908 (1964).

Polanyi, M. (1958, 1962, 1974). *Personal Knowledge: Towards a Post-Critical Philosophy*, ISBN 9780415151498 (1998).

Polanyi, M. (1966). *The Tacit Dimension*, ISBN 9780226672984.

Rosen, R. (1970). "Structural and functional considerations in the modeling of biological organization," Center for Theoretical Biology, State University of New York at Buffalo, 77(25): 1-12, http://rosenenterprises.com/images/RR_Structural_and_Functional_Considerations_in_the_Mode_.pdf.

Rosen, R. (2005). *Life Itself: A Comprehensive Inquiry into the Nature, Origin, and Fabrication of Life*, ISBN 9780231075657.

Roth, S. (2011). "Mathematics and biology: A Kantian view on the history of pattern formation theory," *Development Genes Evolution*, ISSN 0949-944X, 10(221): 255-279.

Swenson, R. (1992). "Autocatakinetics, Yes; Autopoiesis, No: Steps toward a unified theory of evolutionary ordering," *International Journal of General Systems*, ISSN 0308-1079, 21: 207-208.

Turing, A. (1952). "The chemical basis of morphogenesis," *Philosophical Transactions of the Royal Society of London - B*, ISSN 0264-3839, 237: 37-72.

Life's Irreducible Structure

Live mechanisms and information in DNA are boundary
conditions with a sequence of boundaries above them.

Michael Polanyi

If all men were exterminated, this would not affect the laws of inanimate nature. But the production of machines would stop, and not until men arose again could machines be formed once more. Some animals can produce tools, but only men can construct machines; machines are human artifacts, made of inanimate material.

The *Oxford Dictionary* describes a machine as "an apparatus for applying mechanical power, consisting of a number of interrelated parts, each having a definite function." It might be, for example, a machine for sewing or printing. Let us assume that the power driving the machine is built in, and disregard the fact that it has to be renewed from time to time. We can say, then, that the manufacture of a machine consists in cutting suitably shaped parts and fitting them together so that their joint mechanical action should serve a possible human purpose.

The structure of machines and the working of their structure are thus shaped by man, even while their material and the forces that operate them obey the laws of inanimate nature. In constructing a machine and supplying it with power, we harness the laws of nature at work in its material and in its driving force and make them serve our purpose.

This harness is not unbreakable; the structure of the machine, and thus its working, can break down. But this will not affect the forces of inanimate nature on which the operation of the machine relied; it merely releases them from the restriction the machine imposed on them before it broke down.

So the machine as a whole works under the control of two distinct principles. The higher one is the principle of the machine's design, and this harnesses the lower one, which consists in the physical-chemical processes on which the machine relies. We commonly form such a two-leveled structure in conducting an experiment; but there is a difference between constructing a machine and rigging up an experiment. The experimenter imposes restrictions on nature in order to observe its behavior under these restrictions, while the constructor of a machine restricts nature in order to harness its workings. But we may borrow a term from physics and describe both these useful restrictions of nature as the imposing of *boundary conditions* on the laws of physics and chemistry.

Let me enlarge on this. I have exemplified two types of boundaries. In the machine our principal interest lay in the effects of the boundary conditions, while in an experimental setting we are interested in the natural processes controlled by the boundaries. There are many common examples of both types of boundaries. When a saucepan bounds a soup that we are cooking, we are interested in the soup; and, likewise, when we observe a reaction in a test tube, we are studying the reaction, not the test tube. The reverse is true for a game of chess. The strategy of the player imposes boundaries on the several moves, which follow the laws of chess, but our interest lies in the boundaries—that is, in the strategy, not in the several moves as exemplifications of the laws. And similarly, when a sculptor shapes a stone or a painter composes a painting, our interest lies in the boundaries imposed on a material, and not in the material itself.

We can distinguish these two types of boundaries by saying that the first represents a test-tube type of boundary

whereas the second is of the machine type. By shifting our attention, we may sometimes change a boundary from one type to another.

All communications form a machine type of boundary, and these boundaries form a whole hierarchy of consecutive levels of action. A vocabulary sets boundary conditions on the utterance of the spoken voice; a grammar harnesses words to form sentences, and the sentences are shaped into a text which conveys a communication. At all these stages we are interested in the boundaries imposed by a comprehensive restrictive power, rather than in the principles harnessed by them.

Living Mechanisms Are Classed with Machines

From machines we pass to living beings, by remembering that animals move about mechanically and that they have internal organs which perform functions as parts of a machine do—functions which sustain the life of the organism, much as the proper functioning of parts of a machine keeps the machine going. For centuries past, the workings of life have been likened to the working of machines and physiology has been seeking to interpret the organism as a complex network of mechanisms. Organs are, accordingly, defined by their life-preserving functions.

Any coherent part of the organism is indeed puzzling to physiology—and also meaningless to pathology—until the way it benefits the organism is discovered. And I may add that any description of such a system in terms of its physical-chemical topography is meaningless, except for the fact that the description covertly may recall the system's physiological interpretation—much as the topography of a machine is meaningless until we guess how the device works, and for what purpose.

In this light the organism is shown to be, like a machine, a system which works according to two different principles: its structure serves as a boundary condition harnessing the physical-chemical processes by which its organs perform their functions. Thus, this system may be called a system under dual control. Morphogenesis, the process by which the structure of living beings develops, can then be likened to the shaping of a machine which will act as a boundary for the laws of inanimate nature. For just as these laws serve the machine, so they serve also the developed organism.

The author is a former Fellow of Merton College, Oxford, and Emeritus Professor of social studies at the University of Manchester, where he had previously held the Chair of Physical Chemistry. His present address is 22 Upland Park Road, Oxford, England. This article is an expanded version of a paper presented 20 December 1967 at the New York meeting of the AAAS. The first half of the article was anticipated in a paper published in the August 1967 issue of *Chemical and Engineering News*.

1308

A boundary condition is always extraneous to the process which it delimits. In Galileo's experiments on balls rolling down a slope, the angle of the slope was not derived from the laws of mechanics, but was chosen by Galileo. And as this choice of slopes was extraneous to the laws of mechanics, so is the shape and manufacture of test tubes extraneous to the laws of chemistry.

The same thing holds for machine-like boundaries; their structure cannot be defined in terms of the laws which they harness. Nor can a vocabulary determine the content of a text, and so on. Therefore, if the structure of living things is a set of boundary conditions, this structure is extraneous to the laws of physics and chemistry which the organism is harnessing. Thus the morphology of living things transcends the laws of physics and chemistry.

DNA Information Generates

Mechanisms

But the analogy between machine components and live functioning organs is weakened by the fact that the organs are not shaped artificially as the parts of a machine are. It is an advantage, therefore, to find that the morphogenetic process is explained in principle by the transmission of information stored in DNA, interpreted in this sense by Watson and Crick.

A DNA molecule is said to represent a code—that is, a linear sequence of items, the arrangement of which is the information conveyed by the code. In the case of DNA, each item of the series consists of one out of four alternative organic bases (1). Such a code will convey the maximum amount of information if the four organic bases have equal probability of forming any particular item of the series. Any difference in the binding of the four alternative bases, whether at the same point of the series or between two points of the series, will cause the information conveyed by the series to fall below the ideal maximum. The information content of DNA is in fact known to be reduced to some extent by redundancy, but I accept here the assumption of Watson and Crick that this redundancy does not prevent DNA from effectively functioning as a code. I accordingly disregard, for the sake of brevity, the redundancy in the DNA code and talk of it as if it were functioning optimally, with all of its alterna-

tive basic bindings having the same probability of occurrence.

Let us be clear what would happen in the opposite case. Suppose that the actual structure of a DNA molecule were due to the fact that the bindings of its bases were much stronger than the bindings would be for any other distribution of bases, then such a DNA molecule would have no information content. Its codelike character would be effaced by an overwhelming redundancy.

We may note that such is actually the case for an ordinary chemical molecule. Since its orderly structure is due to a maximum of stability, corresponding to a minimum of potential energy, its orderliness lacks the capacity to function as a code. The pattern of atoms forming a crystal is another instance of complex order without appreciable information content.

There is a kind of stability which often opposes the stabilizing force of a potential energy. When a liquid evaporates, this can be understood as the increase of entropy accompanying the dispersion of its particles. One takes this dispersive tendency into account by adding its powers to those of potential energy, but the correction is negligible for cases of deep drops in potential energy or for low temperatures, or for both. We can disregard it, to simplify matters, and say that chemical structures established by the stabilizing powers of chemical bonding have no appreciable information content.

In the light of the current theory of evolution, the codelike structure of DNA must be assumed to have come about by a sequence of chance variations established by natural selection. But this evolutionary aspect is irrelevant here; whatever may be the origin of a DNA configuration, it can function as a code only if its order is not due to the forces of potential energy. It must be as physically indeterminate as the sequence of words is on a printed page. As the arrangement of a printed page is extraneous to the chemistry of the printed page, so is the base sequence in a DNA molecule extraneous to the chemical forces at work in the DNA molecule. It is this physical indeterminacy of the sequence that produces the improbability of occurrence of any particular sequence and thereby enables it to have a meaning—a meaning that has a mathematically determinate information content equal to the numerical improbability of the arrangement.

DNA Acts as a Blueprint

But there remains a fundamental point to be considered. A printed page may be a mere jumble of words, and it has then no information content. So the improbability count gives the *possible*, rather than the *actual*, information content of a page. And this applies also to the information content attributed to a DNA molecule; the sequence of the bases is deemed meaningful only because we assume with Watson and Crick that this arrangement generates the structure of the offspring by endowing it with its own information content.

This brings us at last to the point that I aimed at when I undertook to analyze the information content of DNA: Can the control of morphogenesis by DNA be likened to the designing and shaping of a machine by the engineer? We have seen that physiology interprets the organism as a complex network of mechanisms, and that an organism is—like a machine—a system under dual control. Its structure is that of a boundary condition harnessing the physical-chemical substances within the organism in the service of physiological functions. Thus, in generating an organism, DNA initiates and controls the growth of a mechanism that will work as a boundary condition within a system under dual control.

And I may add that DNA itself is such a system, since every system conveying information is under dual control, for every such system restricts and orders, in the service of conveying its information, extensive resources of particulars that would otherwise be left at random, and thereby acts as a boundary condition. In the case of DNA this boundary condition is a blueprint of the growing organism (2).

We can conclude that in each embryonic cell there is present the duplicate of a DNA molecule having a linear arrangement of its bases—an arrangement which, being independent of the chemical forces within the DNA molecules, conveys a rich amount of meaningful information. And we see that when this information is shaping the growing embryo, it produces in it boundary conditions which, themselves being independent of the physical chemical forces in which they are rooted, control the mechanism of life in the developed organism.

To elucidate this transmission is a major task of biologists today, to which I shall return.

Some Accessory Problems Arise Here

We have seen boundary conditions introducing principles not capable of formulation in terms of physics or chemistry into inanimate artifacts and living things; we have seen them as necessary to an information content in a printed page or in DNA, and as introducing mechanical principles into machines as well as into the mechanisms of life.

Let me add now that boundary conditions of inanimate systems established by the history of the universe are found in the domains of geology, geography, and astronomy, but that these do not form systems of dual control. They resemble in this respect the test-tube type of boundaries of which I spoke above. Hence the existence of dual control in machines and living mechanisms represents a discontinuity between machines and living things on the one hand and inanimate nature on the other hand, so that both machines and living mechanisms are irreducible to the laws of physics and chemistry.

Irreducibility must not be identified with the mere fact that the joining of parts may produce features which are not observed in the separate parts. The sun is a sphere, and its parts are not spheres, nor does the law of gravitation speak of spheres; but mutual gravitational interaction causes the parts of the sun to form a sphere. Such cases of holism are common in physics and chemistry. They are often said to represent a transition to living things, but this is not the case, for they are reducible to the laws of inanimate matter, while living things are not.

But there does exist a rather different continuity between life and inanimate nature. For the beginnings of life do not sharply differ from their purely physical-chemical antecedents. One can reconcile this continuity with the irreducibility of living things by recalling the analogous case of inanimate artifacts. Take the irreducibility of machines; no animal can produce a machine, but some animals can make primitive tools, and their use of these tools may be hardly distinguishable from the mere use of the animal's limbs. Or take a set of sounds conveying information; the set of sounds can be so obscured by noise that its presence is no longer clearly identifiable. We can say, then, that the control exercised by the boundary conditions of a system can be reduced gradually to a vanishing point. The fact that the effect

of a higher principle over a system under dual control can have any value down to zero may allow us also to conceive of the continuous emergence of irreducible principles within the origin of life.

We Can Now Recognize
Additional Irreducible Principles

The irreducibility of machines and printed communications teaches us, also, that the control of a system by irreducible boundary conditions does not *interfere* with the laws of physics and chemistry. A system under dual control relies, in fact, for the operations of its higher principle, on the working of principles of a lower level, such as the laws of physics and chemistry. Irreducible higher principles are *additional* to the laws of physics and chemistry. The principles of mechanical engineering and of communication of information, and the equivalent biological principles, are all additional to the laws of physics and chemistry.

But to assign the rise of such additional controlling principles to a selective process of evolution leaves serious difficulties. The production of boundary conditions in the growing fetus by transmitting to it the information contained in DNA presents a problem. Growth of a blueprint into the complex machinery that it describes seems to require a system of causes not specifiable in terms of physics and chemistry, such causes being additional both to the boundary conditions of DNA and to the morphological structure brought about by DNA.

This missing principle which builds a bodily structure on the lines of an instruction given by DNA may be exemplified by the far-reaching regenerative powers of the embryonic sea urchin, discovered by Driesch, and by Paul Weiss's discovery that completely dispersed embryonic cells will grow, when lumped together, into a fragment of the organ from which they were isolated (3). We see an integrative power at work here, characterized by Spemann and by Paul Weiss as a "field" (4), which guides the growth of embryonic fragments to form the morphological features to which they embryologically belong. These guides of morphogenesis are given a formal expression in Waddington's "epigenetic landscapes" (5). They say graphically that the growth of the embryo is controlled by the gradient of potential shapes, much as the

motion of a heavy body is controlled by the gradient of potential energy.

Remember how Driesch and his supporters fought for recognition that life transcends physics and chemistry, by arguing that the powers of regeneration in the sea urchin embryo were not explicable by a machinelike structure, and how the controversy has continued, along similar lines, between those who insisted that regulative ("equipotential" or "organismic") integration was irreducible to any machinelike mechanism and was therefore irreducible also to the laws of inanimate nature. Now if, as I claim, machines and mechanical processes in living beings are themselves irreducible to physics and chemistry, the situation is changed. If mechanistic and organismic explanations are both equally irreducible to physics and chemistry, the recognition of organismic processes no longer bears the burden of standing alone as evidence for the irreducibility of living things. Once the "field"-like powers guiding regeneration and morphogenesis can be recognized without involving this major issue, I think the evidence for them will be found to be convincing.

There is evidence of irreducible principles, additional to those of morphological mechanisms, in the sentience that we ourselves experience and that we observe indirectly in higher animals. Most biologists set aside these matters as unprofitable considerations. But again, once it is recognized, on other grounds, that life transcends physics and chemistry, there is no reason for suspending recognition of the obvious fact that consciousness is a principle that fundamentally transcends not only physics and chemistry but also the mechanistic principles of living beings.

Biological Hierarchies Consist of
a Series of Boundary Conditions

The theory of boundary conditions recognizes the higher levels of life as forming a hierarchy, each level of which relies for its workings on the principles of the levels below it, even while it itself is irreducible to these lower principles. I shall illustrate the structure of such a hierarchy by showing the way five levels make up a spoken literary composition.

The lowest level is the production of a voice; the second, the utterance of words; the third, the joining of words to make sentences; the fourth, the working of sentences into a style; the fifth, and highest, the composition of the text.

The principles of each level operate under the control of the next-higher level. The voice you produce is shaped into words by a vocabulary; a given vocabulary is shaped into sentences in accordance with a grammar; and the sentences are fitted into a style, which in turn is made to convey the ideas of the composition. Thus each level is subject to dual control: (i) control in accordance with the laws that apply to its elements in themselves, and (ii) control in accordance with the laws of the powers that control the comprehensive entity formed by these elements.

Such multiple control is made possible by the fact that the principles governing the isolated particulars of a lower level leave indeterminate conditions to be controlled by a higher principle. Voice production leaves largely open the combination of sounds into words, which is controlled by a vocabulary. Next, a vocabulary leaves largely open the combination of words to form sentences, which is controlled by grammar, and so on. Consequently, the operations of a higher level cannot be accounted for by the laws governing its particulars on the next-lower level. You cannot derive a vocabulary from phonetics; you cannot derive grammar from a vocabulary; a correct use of grammar does not account for good style; and a good style does not supply the content of a piece of prose.

Living beings comprise a whole sequence of levels forming such a hierarchy. Processes at the lowest level are caused by the forces of inanimate nature, and the higher levels control, throughout, the boundary conditions left open by the laws of inanimate nature. The lowest functions of life are those called vegetative. These vegetative functions, sustaining life at its lowest level, leave open—both in plants and in animals—the higher functions of growth and in animals also leave open the operations of muscular actions. Next, in turn, the principles governing muscular actions in animals leave open the integration of such actions into innate patterns of behavior; and, again, such patterns are open in their turn to be shaped by intelligence, while intelligence itself can be made to serve in man the still higher principles of a responsible choice.

Each level relies for its operations on all the levels below it. Each reduces the scope of the one immediately below it by imposing on it a boundary that

harnesses it to the service of the next-higher level, and this control is transmitted stage by stage, down to the basic inanimate level.

The principles additional to the domain of inanimate nature are the product of an evolution the most primitive stages of which show only vegetative functions. This evolutionary progression is usually described as an increasing complexity and increasing capacity for keeping the state of the body independent of its surroundings. But if we accept, as I do, the view that living beings form a hierarchy in which each higher level represents a distinctive principle that harnesses the level below it (while being itself irreducible to its lower principles), then the evolutionary sequence gains a new and deeper significance. We can recognize then a strictly defined progression, rising from the inanimate level to ever higher additional principles of life.

This is not to say that the higher levels of life are altogether absent in earlier stages of evolution. They may be present in traces long before they become prominent. Evolution may be seen, then, as a progressive intensification of the higher principles of life. This is what we witness in the development of the embryo and of the growing child—processes akin to evolution.

But this hierarchy of principles raises once more a serious difficulty. It seems impossible to imagine that the sequence of higher principles, transcending further at each stage the laws of inanimate nature, is incipiently present in DNA and ready to be transmitted by it to the offspring. The conception of a blueprint fails to account for the transmission of faculties, like consciousness, which no mechanical device can possess. It is as if the faculty of vision were to be made intelligible to a person born blind by a chapter of sense physiology. It appears, then, that DNA *evokes* the ontogenesis of higher levels, rather than *determining* these levels. And it would follow that the emergence of the kind of hierarchy I have defined here can be only evoked, and not determined, by atomic or molecular accidents. However, this question cannot be argued here.

Understanding a Hierarchy
Needs "from-at" Conceptions

I said above that the transcendence of atomism by mechanism is reflected in the fact that the presence of a mech-

anism is not revealed by its physical-chemical topography. We can say the same thing of all higher levels: their description in terms of any lower level does not tell us of their presence. We can generally descend to the components of a lower level by analyzing a higher level, but the opposite process involves an integration of the principles of the lower level, and this integration may be beyond our powers.

In practice this difficulty may be avoided. To take a common example, suppose that we have repeated a particular word, closely attending to the sound we are making, until these sounds have lost their meaning for us; we can recover this meaning promptly by evoking the context in which the word is commonly used. Consecutive acts of analyzing and integrating are in fact generally used for deepening our understanding of complex entities comprising two or more levels.

Yet the strictly logical difference between two consecutive levels remains. You can look at a text in a language you do not understand and see the letters that form it without being aware of their meaning, but you cannot read a text without seeing the letters that convey its meaning. This shows us two different and mutually exclusive ways of being aware of the text. When we look at words without understanding them we are focusing our attention on them, whereas, when we read the words, our attention is directed to their meaning as part of a language. We are aware then of the words only subsidiarily, as we attend to their meaning. So in the first case we are looking at the words, while in the second we are looking *from* them *at their meaning*: the reader of a text has a *from-at* knowledge of the words' meaning, while he has only a *from* awareness of the words he is reading. Should he be able to shift his attention fully toward the words, these would lose their linguistic meaning for him.

Thus a boundary condition which harnesses the principles of a lower level in the service of a new, higher level establishes a semantic relation between the two levels. The higher comprehends the workings of the lower and thus forms the meaning of the lower. And as we ascend a hierarchy of boundaries, we reach to ever higher levels of meaning. Our understanding of the whole hierarchic edifice keeps deepening as we move upward from stage to stage.

The Sequence of Boundaries Bears on Our Scientific Outlook

The recognition of a whole sequence of irreducible principles transforms the logical steps for understanding the universe of living beings. The idea, which comes to us from Galileo and Gassendi, that all manner of things must ultimately be understood in terms of matter in motion is refuted. The spectacle of physical matter forming the basic tangible ground of the universe is found to be almost empty of meaning. The universal topography of atomic particles (with their velocities and forces) which, according to Laplace, offers us a universal knowledge of all things is seen to contain hardly any knowledge that is of interest. The claims made, following the discovery of DNA, to the effect that all study of life could be reduced eventually to molecular biology, have shown once more that the Laplacean idea of universal knowledge is still the theoretical ideal of the natural sciences; current opposition to these declarations has often seemed to confirm this ideal, by defending the study of the whole organism as being only a temporary approach. But now the analysis of the hierarchy of living things shows that to reduce this hierarchy to ultimate particulars is to wipe out our very sight of it. Such analysis proves this ideal to be both false and destructive.

Each separate level of existence is of course interesting in itself and can be studied in itself. Phenomenology has taught this, by showing how to save higher, less tangible levels of experience by not trying to interpret them in terms of the more tangible things in which their existence is rooted. This method was intended to prevent the reduction of man's mental existence to mechanical structures. The results of the method were abundant and are still flowing, but phenomenology left the ideal of exact science untouched and thus failed to secure the exclusion of its claims. Thus, phenomenological studies remained suspended over an abyss of reductionism. Moreover, the relation of the higher principles to the workings of the lowest levels in which they are rooted was lost from sight altogether.

I have mentioned how a hierarchy controlled by a series of boundary principles should be studied. When examining any higher level, we must remain subsidiarily aware of its grounds in lower levels and, turning our attention to the latter, we must continue to see them as bearing on the levels above them. Such alternation of detailing and integrating admittedly leaves open many dangers. Detailing may lead to pedantic excesses, while too-broad integrations may present us with a meandering impressionism. But the principle of stratified relations does offer at least a rational framework for an inquiry into living things and the products of human thought.

I have said that the analytic descent from higher levels to their subsidiaries is usually feasible to some degree, while the integration of items of a lower level so as to predict their possible meaning in a higher context may be beyond the range of our integrative powers. I may add now that the same things may be seen to have a joint meaning when viewed from one point, but to lack this connection when seen from another point. From an airplane we can see the traces of prehistoric sites which, over the centuries, have been unnoticed by people walking over them; indeed, once he has landed, the pilot himself may no longer see these traces.

The relation of mind to body has a similar structure. The mind-body problem arises from the disparity between the experience of a person observing an external object—for example, a cat—and a neurophysiologist observing the bodily mechanism by means of which the person sees the cat. The difference arises from the fact that the person observing the cat has a *from*-knowledge of the bodily responses evoked by the light in his sensory organs, and this *from*-knowledge integrates the joint meaning of these responses to form the sight of the cat, whereas the neurophysiologist, looking at these responses from outside, has only an *at*-knowledge of them, which, as such, is not integrated to form the sight of the cat. This is the same duality that exists between the airman and the pedestrian in interpreting the same traces, and the same that exists between a person who, when reading a written sentence, sees its meaning and another person who, being ignorant of the language, sees only the writing.

Awareness of mind and body confront us, therefore, with two different things. The mind harnesses neurophysiological mechanisms and is not determined by them. Owing to the existence of two kinds of awareness—the focal and the subsidiary—we can now distinguish sharply between the mind as a "from-at" experience and the subsidiaries of this experience, seen focally as a bodily mechanism. We can see then that, though rooted in the body, the mind is free in its actions—exactly as our common sense knows it to be free.

The mind itself includes an ascending sequence of principles. Its appetitive and intellectual workings are transcended by principles of responsibility. Thus the growth of man to his highest levels is seen to take place along a sequence of rising principles. And we see this evolutionary hierarchy built as a sequence of boundaries, each opening the way to higher achievements by harnessing the strata below them, to which they themselves are not reducible. These boundaries control a rising series of relations which we can understand only by being aware of their constituent parts subsidiarily, as bearing on the upper level which they serve.

The recognition of certain basic impossibilities has laid the foundations of some major principles of physics and chemistry; similarly, recognition of the impossibility of understanding living things in terms of physics and chemistry, far from setting limits to our understanding of life, will guide it in the right direction. And even if the demonstration of this impossibility should prove of no great advantage in the pursuit of discovery, such a demonstration would help to draw a truer image of life and man than that given us by the present basic concepts of biology.

Summary

Mechanisms, whether man-made or morphological, are boundary conditions harnessing the laws of inanimate nature, being themselves irreducible to those laws. The pattern of organic bases in DNA which functions as a genetic code is a boundary condition irreducible to physics and chemistry. Further controlling principles of life may be represented as a hierarchy of boundary conditions extending, in the case of man, to consciousness and responsibility.

References and Notes

1. More precisely, each item consists of one out of four alternatives consisting in two positions of two different compound organic bases.
2. The blueprint carried by the DNA molecule of a particular zygote also prescribes individual features of this organism, which contribute to the sources of selective evolution, but I shall set these features aside here.
3. See P. Weiss, *Proc. Nat. Acad. Sci. U.S.* 42, 819 (1956).
4. The "field" concept was first used by Spemann (1921) in describing the organizer; Paul Weiss (1923) introduced it for the study of regeneration and extended it (1926) to include ontogeny. See P. Weiss, *Principles of Development* (Holt, New York, 1939), p. 290.
5. See, for example, C. H. Waddington, *The Strategy of the Genes* (Allen & Unwin, London, 1957), particularly the graphic explanation of "genetic assimilation" on page 167.
6. See, for example, M. Polanyi, *Amer. Psychologist* 23 (Jan. 1968) or ———, *The Tacit Dimension* (Doubleday, New York, 1967).

Forum

Forum

Adjacent Opportunities: Building a Healthy Economy

Ron Schultz
Entrepreneurs4Change, US

I hate being reminded that I'm older than I feel. My friend and colleague, Pat Conaty, did just that recently when I broached the topic of Building Healthy Economies with him. Pat is a brilliant economic social innovator, author, and a person who puts his actions where his ideas are. As he explained "The problem is that just as folks like you and me hit late life (Late life? Ouch) and then have a certain epiphany: we, like previous generations of elders, in our existential quest for hallowed truth, end up rediscovering the wisdom underlying the ancient co-operative wheel of social and economic democracy." He goes on to describe that the positions that he and his coauthor Mike Lewis make in their new book, *The Resilience Imperative*, is that "the social economy and social enterprise sector is not new. Sadly the stone rolls back down the hill. We need to sort this and push mightily to the top. How to mobilize and achieve this non-violently is the question."

OK, I'm older than I feel and my drive to build healthy economies is nothing new. But as Conaty, who actually is 9 days older than I am, goes on to point out "the social economy struggle is as old as the hills, with deep roots in the peaceful struggle of Gerard Winstanley and the Diggers to persuade Cromwell in 1650 to allow commonwealth to be developed by stewarding land for community bene-fit. The Commonwealth was not fulfilled by Cromwell, despite aspirations by the Levellers and the Diggers for economic democracy. Ivan Illich talked brilliantly in his little books about the endless struggle of the vernacular culture against the Romans (in all their ancient and modern costumes)." Well, I'm not that old.

The fact that we have been trying for centuries to enact economic health speaks both to the tenacity of the movement as well as the power of such drivers as manipulation, excess and greed. There are a number of questions that arise for me, however. How do we create businesses, work and ultimately economic systems today that provide enough wealth, as opposed to vast wealth, so that 100% of us have enough? And how do we do so in such a way that we can find meaning and purpose in that labor, while nurturing the resilience that can withstand what happens when things like excess and greed invariably try to bring things down?

These questions, of course, imply a basic definition of what drives a healthy economy.

It begins by nurturing resilience within the system that in turn fosters a limber flexibility to change and adapt, while allowing for the change and adaptation of others. In this instance, the economic system must accommodate all the interactions taking place within it without bursting the boundaries it has established for itself. These interactions are a continual adjustment to the models we have built to achieve, in this case, a healthy economy. The process of dynamic change that takes place in our economic models requires us to be constantly making and remaking the rules that govern appropriate behavior within the system. The challenge is that when we build-in either too few rules or too many, or hold onto rules that are no longer relevant to the current model, the health of the system declines rapidly. Behaviors go awry because without the boundaries of appropriate rules, there is nothing to contain those behaviors that operate against the health of the system. The system is pushed beyond its extremes and it loses its resilience—it bursts, crashes and dies.

To build a healthy economy, the models of our system must reflect what we want. The rules we create to govern behaviors within the system must support that direction, while maintaining the system's ability to accommodate the emergent change that will arise, shifting the model and rules, again and again. Economies are not static—economists perhaps—but not our economies. The resilience of a healthy economy fosters and supports the development of even greater health, creativity and financial well being. Unfortunately, an unhealthy economy, one that fosters greed and excess, has lost its resilience and stifles change, allows for even greater deterioration and disease within the system.

At the forefront of any healthy human system is an ability to accommodate others. In contrast, what we have seen within our economy is a model that is designed to take from the system rather than one built to give back. When we have a system of depletion, there is only one outcome that can arise, and you can bet it isn't good for the greater good. The work that accomplishes this model of depletion, by its very nature, is not designed with the intention of returning anything meaningful or purposeful to the system. It is true that meaning and purpose are subjective terms, relative to the perceiver. But what we see is that meaning and purpose, and real health, comes in our ability to focus on giving back to others and not depleting the system for our own gain. And when we create work as a means of accommodating others, we establish a level of meaning and purpose that, as a participant in that effort, we can find satisfaction and well-being in what we give back. We become focused on the greater good rather than the individual, and we can begin to breakdown the provincialism that says "this is mine and mine alone." Within a healthy economy, mine and ours are interdependent and interconnected. And that really pisses off those who tend to only recognize "mine."

In my forthcoming book, *Creating Good Work: The World's Leading Social Entrepreneurs Show How to Build a Healthy Economy*, to be published by Palgrave Macmillan in February, 2013, over 20 of the world's foremost social innovators describe what work with meaning and purpose looks like. We see the efforts of SecondMuse leading large-scale collaborations that are bringing world-shifting innovations to issues like water, pollution, energy and health. Karen Tse and her colleagues at International Bridges to Justice are training public defenders to end torture in over 30 countries, and they are succeeding. Dorothy Stoneman and her organization, YouthBuild, where she has developed over 250 YouthBuild programs around the country providing education, work, and meaning and purpose to young people who have dropped out of school and whose lives were floundering. Billy Shore and his amazing work at Share Our Strength and its nationwide program to end childhood hunger.

And then comes the cry, "Resilience, meaning and purpose are fine ideas, but alone they don't put bread on the table!" And to a certain degree, that is right. They might teach you how to make a better loaf of bread, but until someone is willing to produce a transaction for that bread, we don't have much of an economy—which brings us to wealth.

It is easy to demonize the vastly wealthy, but if we are to create a truly healthy economy it means that it works for 100% of the community, not just the 1% or 99%. However, in a healthy economy, the question is not how we create vast wealth, rather how we create enough wealth. And coupled with enough wealth is the equally powerful concept of *commonwealth*. This is what we value, own and share together as a community. Ideally, it's also what we buy with our taxes. Pat Conaty was involved in a series of projects that established community land trusts, in which housing was built on commonly-owned land, cutting the price of homeownership in half and making it accessible to members of the commonwealth. Somehow, I don't see the Commonwealths of Virginia and Kentucky lining up to shift their economies into real commonwealth. But to get to a healthier economy, one that benefits the entire population, how do we get to enough is enough? We build it.

Economic health is not a move to an idealized socialist state, but it does mean we have to let go of some long-held and unhealthy beliefs. As Martin Luther

King, Jr. described this transition in his last book before his death, "Communism forgets that life is individual. Capitalism forgets that life is social, and the Kingdom of the brotherhood is found neither in the thesis of Communism nor the antithesis of Capitalism, but in a higher synthesis. It is found in a higher synthesis that combines the truth of both."

Perhaps Pat Conaty is right, my desire for a healthy economy—where exploitation for the sake of more is replaced by enough for everyone, where there's enough work available that uplifts and inspires one's life, and there's a chance to create and affect real change in the world so we can all live lives that are not overcome by despair, disappointment and dissatisfaction—is just a dream of yet another old man. If that's the case, then let me defy my age and dedicate the rest of my work life so that I can inspire and motivate the dreams of young men and women everywhere, and we won't have to continue making the same mistakes millennia after millennia. And we can declare, enough is enough, and let the economic health and well-being of the people who live within our commonwealth be what makes us great.

Ron Schultz is the author of the forthcoming book, *Creating Good Work: The World's Leading Social Entrepreneurs Show How to Build a Healthy Economy*, Palgrave Macmillan, 2013, ISBN 9780230372030). Enter code XP356ED and save 20% when you order at Palgrave.com.

Calling notices

Calling Notices and Announcements

Contacts

Program Chairs:
Robert Eberlein
astuteSD
Ignacio J. Martínez-Moyano
*Argonne National Laboratory
and The University of Chicago*

Workshop Chairs:
Jack B. Homer
Homer Consulting
Hazhir Rahmandad
Virginia Tech

Conference Manager:
Roberta L. Spencer
*Executive Director
System Dynamics Society*

31st International Conference of the System Dynamics Society
Creating the Future from Within

Cambridge, Massachusetts, USA July 21 - 25, 2013
Call for Papers, Presentations, Workshops, and Sessions

The 2013 program will emphasize the role of endogeneity in shaping system performance and in creating successful interventions. Studies that communicate the relationship between structure and behavior in novel and broadly understandable ways are especially encouraged.

We welcome all research and documented consulting activities in system dynamics including applications of the methodology to solve real world problems, new technical and software developments, and productive integration of complementary methodologies in order to create new solutions. The topics to be addressed include:

- Business (profitability, marketing, competitive dynamics, product launches, project dynamics, and accounting)
- Economics (macroeconomics, microeconomics, trade, business regulation, economic development, economic policy, insurance and risk management)
- Environment (climate change, pollution, environmental regulation, and ecology)
- Health (health policy, health services research, population health and physiology)
- Human Behavior (families, communities, organizations, culture, and society including individual and social psychology)
- Information and Knowledge (knowledge management, information systems, social network analysis, research & development, invention and innovation)
- Learning and Teaching (pedagogy, learning experiments, curriculum development, workshop design and interactive activities)
- Methodology (modeling and simulation including model development, model analysis, validation, graphical presentation formats, and computational techniques)
- Operations (capacity management, quality control, operations management, supply chains, workflow, queuing, and workforce planning)
- Public Policy (governance, social welfare, equity, justice, political science, urban dynamics, infrastructure, and transportation)
- Resources (energy, electricity, fuels, food, metals and other renewable and nonrenewable resources)
- Security (defense, conflict, military, insurgency and counterinsurgency, social unrest, disaster management, crime, policing and incarceration)
- Stakeholder Engagement: (group model building, facilitation, facilitated modeling, games and management flight simulators with emphasis on assessing the impact)
- Strategy (strategic thinking, goal definition, and outcome measures for public and private organizations)

Overview Leading academics, aspiring students, seasoned consultants, new practitioners, and representatives from education, industry, and government will come together at the conference, giving you the opportunity to meet the most active people in the field. The conference program will consist of invited and contributed presentations, poster discussions, and workshops. There will also be panel discussions, special interest group sessions, student colloquia, modeling assistance workshops, cultural events, vendor displays, exhibits, demonstrations, meetings, and other related activities. The schedule will provide ample time for relaxed social and professional interaction.

Venue Cambridge is often referred to as "Boston's Left Bank." The conference hotel, the Hyatt Regency Cambridge, is situated along the scenic Charles River near Massachusetts Institute of Technology (MIT).

Deadline Submissions are due by March 19, 2013.

conference.systemdynamics.org   conference@systemdynamics.org